Advanced Textbooks in Control and Signal Processing

More information about this series at http://www.springer.com/series/4045

László Keviczky · Ruth Bars
Jenő Hetthéssy · Csilla Bányász

Control Engineering: MATLAB Exercises

László Keviczky
Institute for Computer Science
and Control
Hungarian Academy of Sciences
Budapest, Hungary

Jenő Hetthéssy
Department of Automation
and Applied Informatics
Budapest University of Technology
and Economics
Budapest, Hungary

Ruth Bars
Department of Automation
and Applied Informatics
Budapest University of Technology
and Economics
Budapest, Hungary

Csilla Bányász
Institute for Computer Science
and Control
Hungarian Academy of Sciences
Budapest, Hungary

ISSN 1439-2232 ISSN 2510-3814 (electronic)
Advanced Textbooks in Control and Signal Processing
ISBN 978-981-13-4122-9 ISBN 978-981-10-8321-1 (eBook)
https://doi.org/10.1007/978-981-10-8321-1

This Springer imprint is published by the registered company Springer Nature Singapore Pte Ltd
The registered company address is: 152 Beach Road, #21-01/04 Gateway East, Singapore 189721, Singapore

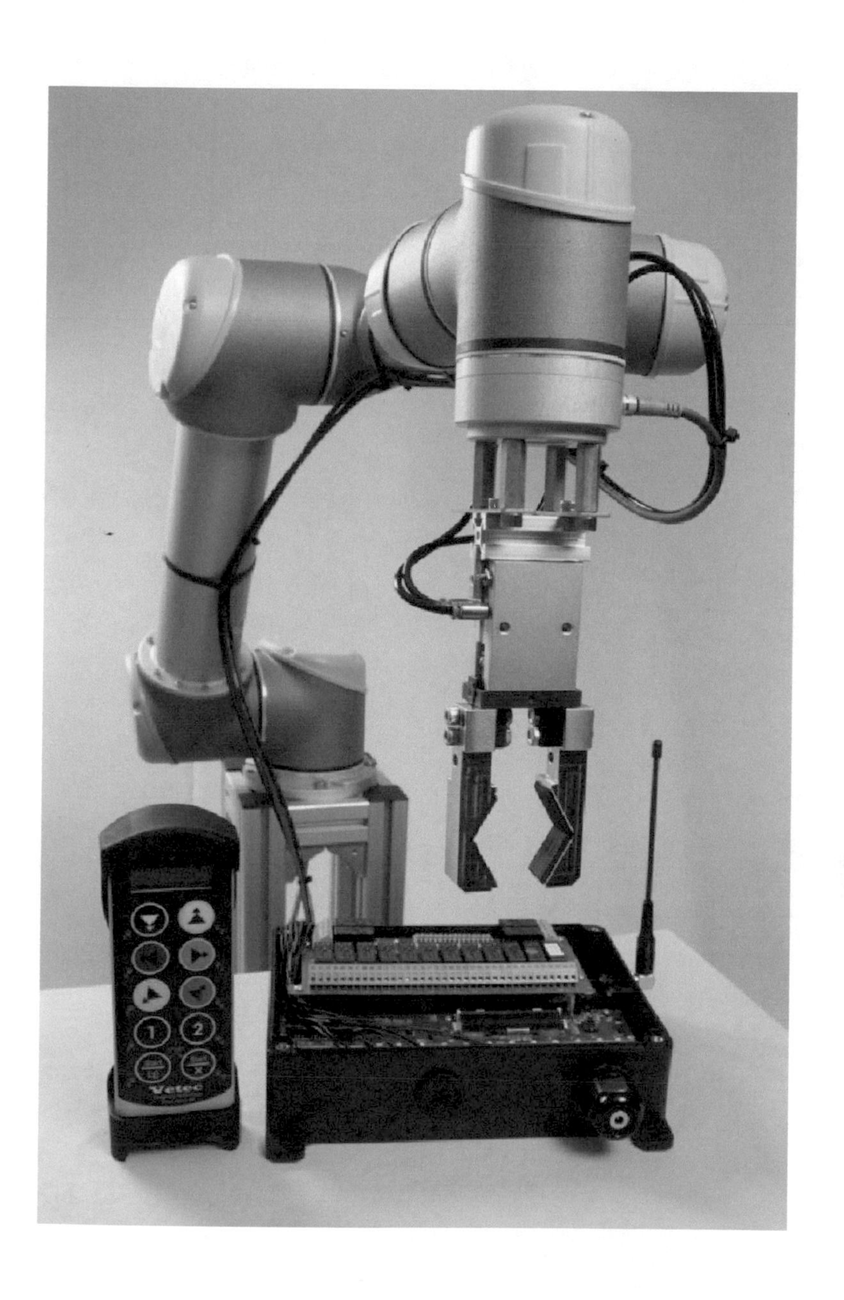
Vetec

Frigyes Csáki
(1921–1977)

This textbook is devoted to the memory of Frigyes Csáki, who was the first professor of control in Hungary

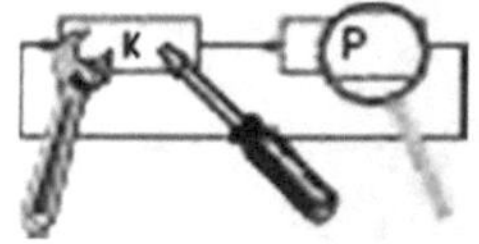

Preface

This book is intended to aid students in their study of MATLAB™/SIMULINK™ for use in solving control problems. Specifically, 16 labs for an introductory control course have been developed at the Department of Automation and Applied Informatics, Budapest University of Technology and Economics. This book is a collection of these labs. This exercise book is a supplement to the textbook "Control Engineering," by László Keviczky, Ruth Bars, Jenő Hetthéssy, and Csilla Bányász [1], which is used in the control course. Each chapter of this exercise book is related to the corresponding chapter of the textbook.

The importance of accompanying textbooks by labs using CAD software was recognized decades ago at the department. At that time, a set of FORTRAN libraries supported the instruction both in control systems analysis and design. We still believe that learning control theory is best motivated by applications and simulations rather than by concepts alone. In fact, the use of MATLAB™ allows a lot of theoretical concepts to be easily implemented. If students can immediately show for themselves how certain concepts work in practice, they will go back to the theoretical considerations with greater confidence and an improved ability to move to the next field to study. Well, feedback is around us, anyway.

The problems discussed in this book are limited to linear, time-invariant control systems. Both continuous-time and discrete-time systems are considered, with deterministic inputs.

MATLAB™/SIMULINK™ is useful only for those students, who master the tools offered. Though the application of MATLAB™ commands is simple and straightforward, a systematic introduction together with control-related examples is a must in our opinion. Time should be devoted to practicing fundamental MATLAB™ facilities, alternative command sequences, and visualization capabilities. An introductory lab is devoted to demonstrating the availability and power of MATLAB™ in this respect.

Frequency functions and transfer functions form essential tools in classical control theory. Interestingly enough, the frequency domain considerations gave a remarkable impetus to the postmodern control era, as well. Three labs have been devoted to discussing fundamental analysis of continuous-time systems including feedback and stability. As far as controller synthesis is concerned, three labs treat YOULA-parameterized control design, as well as *PID* compensation and series compensation for processes with dead time have also been elaborated. The case of controlling unstable processes is also involved. The theoretical discussion of state-space representations is supported by two labs offering a gentle introduction to the subject, as well as demonstrating the efficient algorithms and MATLAB™ commands available for state variable feedback.

These days, controllers are implemented as digital controllers. As most of the processes to be controlled are continuous time in nature, digital control needs additional tools to cover sampled data systems. Just to support the development of a proper view of discrete-time systems, an introductory lab has been added to this topic. Two labs are devoted to discrete controller design. One of them shows controller design using the YOULA parameterization, and the design of a SMITH predictor as well as a deadbeat control as special cases of YOULA parameterization. The second lab discusses discrete-time *PID* controller design. State feedback control for discrete systems is also provided in a lab devoted to this topic.

Two labs deal with the polynomial design method for the compensation of unstable processes, both for the continuous and the discrete case.

In the last lab, the modelling and simulation of a heating process provide a case study.

Each lab is introduced by summarizing the basic concepts and definitions of the topic discussed. The MATLAB™-related functions are discussed in detail. Labs have been designed to be accomplished within a two-hour period, each. Solved examples and reinforcement problems are intended to foster a better understanding. Examples range from simple drills just to demonstrate the MATLAB™ commands to more complex problems, and in most cases a short evaluation completes the lab. It is supposed that the reader writes and runs the codes and evaluates the results. In some cases, the plots are not included in the book, but the evaluation is given, supposed that the reader, after having run the codes, sees the figures.

It is to be emphasized that this set of labs is not a substitute for a textbook in any respect. The textbook of our introductory control course intends to give a deep and comprehensive treatment of control-related subjects. The labs in this book are intended to serve as pedagogical tools offering the student a chance for active learning and experimenting. The present set of labs have been employed in instruction for several semesters.

The authors hope that through active problem solving the students will understand better the control principles and get practice how to apply them in analysis and design of control systems.

Contents

Notations

Transfer functions of continuous-time systems	H (or P)
Transfer functions of discrete-time systems	G
Controller transfer function	C
Process transfer function	P
Discrete-time process pulse transfer function	G (or P_d)
Sensitivity function	S
Complementary sensitivity function	T
Transfer function of an open control loop	L
Gain of a control loop	K
Transfer coefficient of a control loop	k
Youla parameter	Q
Continuous time	(t)
Discrete time	$[k]$
Laplace transformation	$\mathcal{L}\{\ldots\}$
Fourier transformation	$\mathcal{F}\{\ldots\}$
z-transformation	$\mathcal{Z}\{\ldots\}$
Complex variable ($\mathscr{L}$ transformation)	s
Complex variable ($\mathcal{Z}$ transformation)	z
Reference signal	r (or y_r)
Controlled variable	y
Error signal	e
Actuating signal (or output of the regulator)	u
Input noise	y_{ni}
Output noise	y_n (or y_{no})
Measurement noise	y_z
Vector	$\boldsymbol{a}, \boldsymbol{b}, \boldsymbol{c}, \ldots$
Row vector	$\boldsymbol{a}^T, \boldsymbol{b}^T, \boldsymbol{c}^T, \ldots$
Matrix	$\boldsymbol{A}, \boldsymbol{B}, \boldsymbol{C}$
Transpose of a matrix	$\boldsymbol{A}^T$
Adjunct of a matrix	$\mathbf{adj}\,(\boldsymbol{A})$

Determinant of a matrix	$\det(\boldsymbol{A})$ (or $\lvert\mathbf{A}\rvert$)
State variable	$\boldsymbol{x}$
Parameters of the state equation (continuous)	$\boldsymbol{A}, \boldsymbol{b}, \boldsymbol{c}, d$
Parameters of the state equation (discrete)	$\boldsymbol{F}, \boldsymbol{g}, \boldsymbol{h}, d$ (or $\boldsymbol{F}, \boldsymbol{g}, \boldsymbol{c}, d$)
Diagonal matrix	$\mathbf{diag}[a_{11}, a_{22}, \ldots, a_{nn}]$
Unit matrix	$\boldsymbol{I} = \mathbf{diag}[1, 1, \ldots, 1]$
Sampling time	T_s
Dead time (continuous)	T_d
Time delay (discrete)	d
Additional time delay	T_h
Step response function	$v(t)$
Weighting function	$w(t)$
Frequency	ω
Crossover (cutoff) frequency	ω_c
Frequency spectrum of a continuous signal	$F(j\omega)$
Frequency spectrum of a sampled signal series	$F^*(j\omega)$
Frequency spectrum of a discrete-time model	$G(j\omega)$ (or $P_d(j\omega)$)
Polynomials	$\mathcal{A}, \mathcal{B}, \mathcal{C}, \mathcal{D}, \mathcal{G}, \mathcal{F}, \mathcal{R}, \mathcal{X}, \mathcal{Y}, \mathcal{V}$
Degree of a polynomial	$\deg\{\boldsymbol{A}\}$
Characteristic equation	$\mathcal{A}(s) = 0$
Limit of the control output	$\mathbb{U}$
Gradient vector	$\mathbf{grad}[f(\boldsymbol{x})]$
For all ω	$\forall\omega$
Angle of a complex number or function	$\angle$ (or $\mathrm{arc}(\ldots)$)
Exponential function	$e^{(\ldots)}$ (or $\exp(\ldots)$)
Natural logarithm	$\ln(\ldots)$
Base 10 logarithm	$\lg(\ldots)$
Expected value	$E\{\ldots\}$
Probability limit value	$\mathrm{plim}\{\ldots\}$
Matrix exponential	$e^{\mathbf{A}}$
Matrix logarithm	$\ln(\mathbf{A})$
Continuous time	CT
Discrete time	DT
Step response equivalent	*SRE*
Partial fractional expansion	*PFE*

Chapter 1
Introduction to MATLAB

MATLAB™ is an interactive environment for scientific and engineering calculations, simulations, and data visualization. MATLAB™ provides a powerful platform to solve mathematical and engineering problems related to matrix algebra, differential equations, etc. The basic set of MATLAB™ operations can be extended by *toolboxes*. A toolbox is a function library developed to support calculations in a specific subject area. Such special subject areas include signal processing (*Signal Processing Toolbox*), control engineering (*Control System Toolbox*), image processing (*Image Processing Toolbox*), identification (*Identification Toolbox*), the application of neural networks (*Neural Network Toolbox*), etc. The graphical interface of SIMULINK™ provides possibilities for modelling and simulating processes.

MATLAB™ works as an interpreter: it executes the commands row by row. This mode generally results in slow operation. Programs written in MATLAB™ can be accelerated by using matrix operations. In this case there is no need to write cycles: the inner code of MATLAB™ realizes these operations. As matrix operations in MATLAB™ are executed in optimized machine code, the runtime of these programs is similar to that written in other programming languages (e.g. C++); in the case of big matrices, the runtime is even shorter.

The commands can be MATLAB™ functions or so-called *script* files. An *m*–file is a simple text file containing a sequence of MATLAB™ commands and it has the *.m* extension. This series of commands can be executed by writing the name of the file (without the extension). From the MATLAB™ *m*-files, C and C++ files or function libraries (″.lib″, ″.dll″) can also be created using the MATLAB™ translator.

L. Keviczky et al., *Control Engineering: MATLAB Exercises*,
Advanced Textbooks in Control and Signal Processing,
https://doi.org/10.1007/978-981-10-8321-1_1

1.1 Basic Operation of MATLAB™

The goal of this introduction is to enable the newcomer to use MATLAB™ as quickly as possible. However, for detailed descriptions the user should consult the MATLAB™ manuals. They can be found in electronic form in the '*matlab/help*' directory. Also, on-line help is at the MATLAB™ user's disposal.

```
helpdesk
```

The *help* command displays information about any command. For example:

```
help sqrt
```

A script file of MATLAB™ commands can be created. This is a text file with the extension "*.m*". This script file can be used as a new command (without the extension).

Variable names: The maximum length is 31 characters (letters, numbers and underscore). The first character must be a letter. Lower and upper cases are distinguished. Every variable is treated as a matrix. A scalar variable is a 1 by 1 matrix.

1.1.1 Data Entry

If data entry or any other statement/operation is not terminated by a semicolon, the result of the statement will always be displayed. MATLAB™ can use several types of variables. The type declaration is automatic.

Integer:

```
k=2
```

If the command is ended by a semicolon, then the result is not shown on the display, e.g.:

```
J=-4;
```

Real:

```
s=3.6
F2=-12.6e-5
```

Complex:

```
z=3+4*i
r=5*exp(i*pi/3)
```

Although **i=sqrt(-1)** is predefined, you may want to denote the unit imaginary vector by another variable. You are allowed to do so, e.g. simply type

```
j=sqrt(-1)
```

Vectors:

```
x=[1, 2, 3]  % row vector, its elements are separated by commas or spaces
q=[4; 5; 6]  % column vector, its elements are separated by semicolons
```

A column vector can be formed from a raw vector by transposition

```
v=[4, 5, 6]'  % the same as q
```

Remark be careful when using the transpose operation! For complex variables it results in the complex conjugate:

```
i'
     0 - 1.000i
```

Matrices:

```
A=[7, 8, 9; 5, 6, 7]
```

Here **A** is a 2 x 3 matrix, MATRIX = [row1; row2; ...; rowN];

Special vectors and matrices:

```
u=1:3;        % generates u = [1 2 3] as a row vector, » u = start : stop
w=1:2:10;     % generates w = [1 3 5 7 9],     » w = start : increment : stop
E=eye(4)
    E=
         1    0    0    0
         0    1    0    0
         0    0    1    0
         0    0    0    1
B=eye(3,4)
    B=
              1    0    0    0
              0    1    0    0
              0    0    1    0
C=zeros(2,4)
    C=
              0    0    0    0
              0    0    0    0
```

```
D=ones(3,5)
    D=
              1    1    1    1    1
              1    1    1    1    1
              1    1    1    1    1
```

Variable values:

Typing the name of a variable displays its value:

```
A
   A=
         7 8 9
         5 6 7
A(2, 3)
    ans =
          7
```

The first index is the row number and the second index is the column number. The answer is stored in the `ans` variable.

Changing one single value in **v** results in the printing of the entire vector **v**, unless printing is suppressed by a semicolon:

```
v(2)= -6
    v=
        4
       -6
        6
```

Subscripting: A colon (:) can be used to access multiple elements of a matrix. It can be used in several ways for accessing and setting matrix elements.

Start index : end index—means a part of the matrix

: a colon in the index means all the elements in a row or in a column

For vectors:	v=[v(1) v(2) . . . v(N)]
For matrices:	M=[M(1,1)...M(1,m); M(2,1)...M(2,m); ... ; M(n,1)...M(n,m)]

Assume B is an 8 × 8 matrix, Then

B(1:5,3)	is a column vector, **[B(1,3); B(2,3); B(3,3); B(4,3); B(5,3)]**
B(2:3,4:5)	is a matrix **[B(2,4) B(2,5); B(3,4) B(3,5)]**
B(:,3)	assigns all the elements of the third column of **B**
B(2,:)	assigns the second row of **B**
B(1:3,:)	assigns the first three rows of **B**
A(2,1:2)	% second row of matrix A, with its first and second elements
A(:,2)	% all the elements in the second column

1.1.2 Workspace

The used variables are stored in a memory area called the *workspace*. The workspace can be displayed by the following commands:

```
who
whos            % displays also the size of the variables
```

The size of the variables can be displayed by the commands length and size.

For vectors:

```
lng=length(v)
  lng=
        3
```

For matrices and vectors:

```
[m,n]=size(A)
  m=
     2  % number of rows

  n=

     3  % number of columns
```

The workspace can be saved, loaded and cleared:

save	% saves the workspace to the default *matlab.mat* file.
save filename.mat	% saves the workspace to filename.mat file.
clear	% clears the workspace, deletes all the variables.
load	% loads the default matlab.mat file from the workspace.
load filename.mat	% loads the filename.mat file from the workspace.

1.1.3 Arithmetic Operations

Addition and subtraction:

```
A=[1  2;  3  4];
B=A';
C=A+B;
          C=
              2  5
              5  8
D=A-B
          D=
              0  -1
              1  0
x=[-1 0 2]';
y=x-1                    % Observe that all entries are affected!
     y=
          -2
          -1
           1
```

Multiplication:

Vector by scalar:

```
2*x
       ans=
             -2
              0
              4
```

Matrix by scalar:

```
3*A
       ans=
```

```
    3    6
    9   12
```

Inner (scalar) product:

```
s=x'*y
    s=
        4
    y'*x
    ans=
        4
```

Outer product:

```
M=x*y'
    M=
        2    1    -1
        0    0     0
        -4   -2    2
y*x'
    ans=
        2    0    -4
        1    0    -2
        -1   0     2
```

Matrix by vector:

```
b=M*x
    b=
      -4
       0
       8
```

Division:

```
c/2
```

For matrices: B/A corresponds to B^*A^{-1}; A\B corresponds to $A^{-1}*B$.

Powers: A^p, where A is a square matrix and p is a real constant, e.g.: the inverse of A:
A^(-1), or equivalently one can use the command inv(A).

1.1.4 Manipulations of Complex Numbers

```
c=4+2i
     c =
          4.0000 + 2.0000i
real(c)
     ans =
          4
imag(c)
     ans =
          2
abs(c)
     ans =
          4.4721
angle(c)  % the result is in radian
     ans =
          0.4636
```

To get the phase in degrees

```
angle(c)*180/pi
      ans =
         26.5651
```

1.1.5 Array Operations Element-by-Element

Element by element (.*) operations on arithmetic arrays constitute an important class of operations. To indicate an array operation to be executed elementwise the operator should be preceded by a point: a.*b = [a(1)*b(1), a(2)*b(2), ..., a(n)*b(n)]. The sizes of the variables must be the same.

Example

```
a=[2  4  6]
b=[5  3  1]
a.*b
       ans=
              10   12   6
```

The command can be used for different operations, e.g. for division and powers: ./, .^

1.1.6 *Elementary Mathematical Functions*

(Use the on-line help for details and additional items)

abs	absolute value or magnitude of a complex number
sqrt	square root
real	real part
imag	imaginary part
conj	complex conjugate
round	round to nearest integer
fix	round towards zero
floor	round towards -infinity
ceil	round towards +infinity
sign	signum function
rem	remainder
sin	sine
cos	cosine
tan	tangent
asin	arcsine
acos	arccosine
atan	arctangent
atan2	four quadrant arctangent
sinh	hyperbolic sine
cosh	hyperbolic cosine
tanh	hyperbolic tangent
exp	exponential base *e*
log	natural logarithm
log10	log base 10
bessel	BESSEL function
rat	rational approximation
expm	matrix exponential
logm	matrix logarithm
sqrtm	matrix square root

For example:

```
help sqrt
g=sqrt(2)
```

1.1.7 Cell Array Data Type

A *cell array* is a matrix whose elements are also matrices. A *cell array* can be given e.g. as follows:

```
ca={1, [1,2],[1,2,3]}
    ca = [1]   [1x2 double]   [1x3 double]
ca{2}
     ans = 1    2
```

1.1.8 Graphics Output

The most basic graphics command is plot.

```
plot(2,3)              % plots the point given by the coordinates x=2, y=3.
```

Multiple points can be plotted by storing the coordinate values in vectors.

```
x=[1,2,3]
y=[0,2,1]
plot(x,y)              % the points are connected with a line
plot(x,y,'*')          % only the points are plotted
```

This method can plot quite sophisticated curves, too.

Example

```
t=0:0.05:4*pi;
y=sin(t);
plot(t,y)
title('Sine function'),
xlabel('Time'),ylabel('sin(t)'),grid on;
```

where the title, xlabel, ylabel and grid commands are optional.

Plotting more curves in the same coordinate system:

```
y1=3*sin(2*t);
plot(t,y,'r',t,y1,'b'); % r-red, b-blue
```

The typeface and colour for plotting (optional) can be given as follows: » plot(t, y, '@#'), where '@' means line type:

–	solid
– –	dashed
:	dotted
.	point
+	plus
*	star
o	circle
x	x-mark

and '#' means colour as follows:

r	red
g	green
b	blue
w	white
y	yellow

1.1.9 Polynomials

To define a polynomial, e.g. $\mathcal{P}(x) = 2x^5 - 3x^4 + 5x^3 - x^2 - 10x$ simply introduce a vector containing the coefficients of the polynomial:

```
p=[2 -3 5 -1 -10 0]
```

The roots of the polynomial, i.e. the solutions of the equation $\mathcal{P}(x) = 0$ can be calculated by the command roots.

```
xi=roots(p)
  xi =
      0
      0.4756 + 1.7910i
      0.4756 - 1.7910i
      1.5119
     -0.9630
```

It can be seen that this polynomial has three real and two complex roots. Complex roots always appear in conjugate pairs.

From the roots $p_1, p_2, \ldots, p_n$ of a polynomial, the coefficients of the polynomial $\mathcal{P}(x) = (x - p_1)(x - p_2)\ldots(x - p_n)$ can be calculated by

```
p1=poly(xi)
    p1 =
        1.0000 -1.5000 2.5000 -0.5000 -5.0000 0
```

The command poly results in a polynomial with a leading coefficient of 1, therefore to get the original polynomial we have to multiply by 2.

```
2*p1
```

It can be seen that we have obtained the original polynomial.

Let us graph the polynomial $\mathcal{P}(x)$ in the region [−1.5,2].

```
x=[-1.5:0.01:2]
y=p(1)*x.^5+p(2)*x.^4+p(3)*x.^3+p(4)*x.^2+p(5)*x+p(6);
plot(x,y),grid
```

In Fig. 1.1, it can be seen that the crosspoints of the function with the x axis coincide with the real roots.

Consider now the following matrix:

```
M=[3 5; 7 -1]
```

The eigenvalues of $\boldsymbol{M}$ can be computed by the command eig.

```
e=eig(M)
    e =
        7.2450
       -5.2450
```

Taking the above values as the roots of a polynomial, that polynomial can be calculated by the command poly

```
poly(e)
    ans =
        1 -2 -38
```

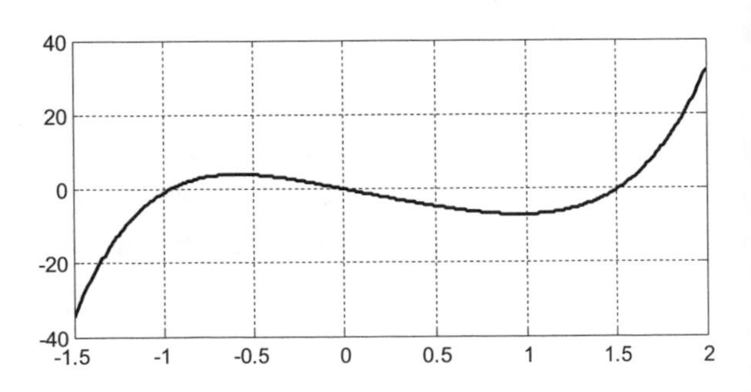

Fig. 1.1 Plotting a polynomial

The above polynomial $x^2 - 2x - 38$ is the characteristic polynomial of $\boldsymbol{M}$, which is defined as $\det(x\boldsymbol{I} - \boldsymbol{M})$. The characteristic polynomial can be directly calculated from $\boldsymbol{M}$:

```
poly(M)
    ans =
          1 -2 -38
```

As can be seen, sometimes commands can be called in different ways. MATLAB™ help of the particular command will provide all the possibilities for how to call it.

1.1.10 Writing MATLAB™ Programs

In the simplest mode of using MATLAB™, we write the commands to be executed in the *command window* of MATLAB™. For solving more complex tasks, this is a long procedure difficult to implement. MATLAB™ provides several possibilities for writing programs. There are two ways to produce programs: in the form of a *script* file or in the form of a *function* file. These programs are simple text files which contain MATLAB™ commands as rows. The extension of the files is *.m*, therefore they are called m-files. The m-files can be written in any text editor, but it is preferable to use the text editor of MATLAB, as it provides several helps for formatting and finding and correcting errors.

A script file program contains MATLAB commands. It can be run in several ways. We can write the name of the file without an extension in the command window, then by the command *Run* from the *menu* or by pressing button F5 (which also saves the modified file) we can run the file. The variables used in the script file appear in the global *workspace*, so their values can be seen from the command window. As an example, let us write a simple script file and run it. Let us create it: File–>New-Script (or m-file).

```
a=2
bscript=2*a+1
```

Let us save the file with the name *myscript.m*. Let us run it. If we run the m-file from the MATLAB™ command window, then in the *Current Folder* window we have to set the place, where has the m-file is to be saved (the *path* can also be set). We can see the result on the screen. The command *whos* ckecks that the variable bscript has been created.

A function can be created using the *function* m-file. The function may have one or more input and output parameters, and it uses local variables. The first row of the m-file contains the key word *function*. Let us create a function m-file.

```
function y=myfunction(x)
bfunction=3*x
y=bfunction+2
```

Let us save the file with the name *myfunction.m*. Call the function from the MATLAB™ command editor:

```
myfunction(4)
```

Using the command *whos* we can check that the global *workspace* does not contain the local variable bfunction. The functions can be embedded in each other in a file, but only the upper function block can be reached from outside.

From the menu, a number of debug devices are available to facilitate programming (Breakpoint, Step, Continue).

1.2 Introduction to the MATLAB™ *Control System Toolbox*

The *Control System Toolbox* extends the toolset of MATLAB™ so as to carry out the analysis, modeling, and design of control systems. The toolbox provides a repertory of algorithms and functions for these purposes, written mainly in the m-file format.

1.2.1 The Use of Functions of the Control System Toolbox

Consider a single-input, single-output (*SISO*), continuous-time, linear, time invariant (*LTI*) system defined by its transfer function (Fig. 1.2.):

Using MATLAB™, we can calculate the step response of a system with the transfer function $H(s) = \frac{2}{s^2+2s+4} = \frac{num}{den}$. The step response is defined as the output $y(t)$ of the system applying a unit step function input $u(t) = 1(t)$ assuming zero initial conditions.

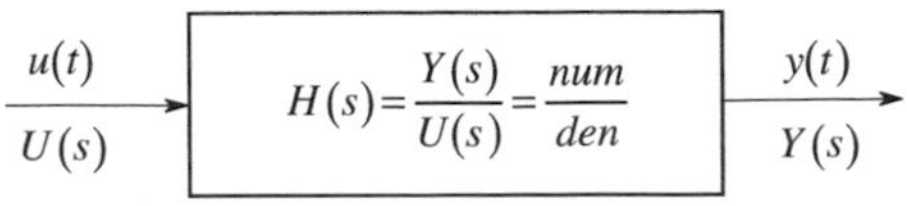

Fig. 1.2 An LTI system defined by its transfer function

The transfer function can be defined in MATLAB™ by its numerator and denominator as polynomials: $num = 2$, $den = s^2 + 2s + 4$. The polynomials are given by their coefficients put in a vector in descending order of s:

```
num=2
den=[1 2 4]
```

The step response can be displayed directly by the MATLAB™ step command (Fig. 1.3.):

```
step(num,den);
```

Note that it is equivalent to use the compact form

```
step(2,[1 2 4]);
```

The time scale is automatically selected by MATLAB™.

Expanding the above command by a left-hand side argument it is possible to store the values of the step response function (the output signal, the state variables and the time vector) in an array:

```
[y,x,t]=step(num,den)
```

or more simply if only the values of the output signal are requested:

```
y=step(num,den)
```

It has to be mentioned that when calling MATLAB™ functions the number of arguments in both sides may vary. The left-hand side output variables in the first activation of the step function are the output variables. y is the output of the step response, t gives the time points where it has been calculated, while x provides the so-called inner or state variables. Let us observe that in this case the output signal is not plotted. The values stored in a variable can be displayed by typing the name of the variable.

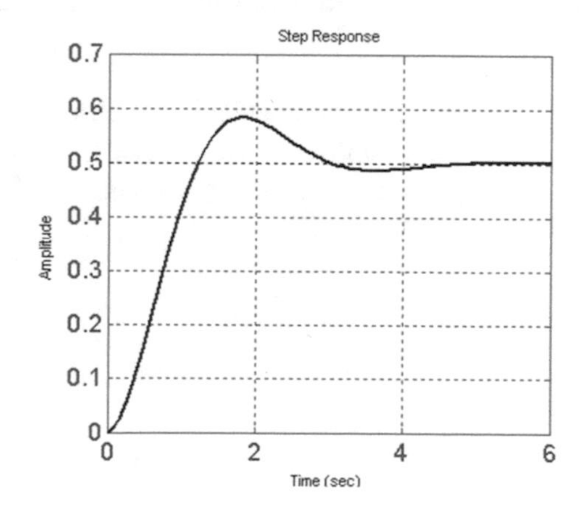

Fig. 1.3 Step response

```
y
```

The result is a column vector whose elements are the calculated values at the sampled points of the step response function. It can be seen from the figure that MATLAB™ has chosen the time interval $0 \le t \le 6$ based on the system's dynamical properties (zeros, poles). The sampling time applied by MATLAB™ can be calculated from the time interval and the size of the vector y:

```
n=length(y)
        n = 109
T=6/n
        T = 0.055
```

The calculated sampling time is thus $T{=}6/109{=}\ 0.055$ s.

The **help** command shows further possible forms of using the step command:

```
help step
```

It can be seen, then, that there are other ways to use the **step** command. E.g. if the time interval $0 \le t \le 10$ and the sampling time $T = 0.1$ are explicitly selected by

```
t=0:0.1:10
```

the following form can be employed:

```
y=step(num,den,t)
```

The output vector can now be displayed with the plot command:

```
plot(t,y);
```

or adding the grid option to support the easy reading of the plot

```
plot(t,y),grid on;
```

As far as the visualization is concerned, the plot command uses linear interpolation between the calculated samples. To avoid this interpolation, the command

```
plot(t,y,'.');
```

displays only the calculated samples.

The obtained y vector can be used for further calculations. The maximum of the step response (more precisely the largest calculated sample) can be determined by command max:

```
ym=max(y)
    ym = 0.5815
```

The steady state value of the step response is obtained by the dcgain command:

```
ys=dcgain(num,den)
    ys = 0.5
```

and the percentage overshoot of the output is

```
yovrsht=(ym-ys)/ys*100
    yovrsht = 16.2971
```

1.2.2 LTI Model Structures (sys Forms)

In order to simplify the commands the Control System Toolbox can also use data-structures. There are three basic forms to describe linear time-invariant (*LTI*) systems in MATLAB™:

Transfer function form: $H_{\mathrm{tf}}(s) = \dfrac{s^m + b_{m-1}s^{m-1} + \ldots + b_2 s^2 + b_1 s + b_0}{a_n s^n + a_{n-1}s^{n-1} + \ldots + a_2 s^2 + a_1 s + a_0} = \dfrac{2}{s^2+3s+2}$

Zero-pole-gain form: $H_{\mathrm{zpk}}(s) = k\dfrac{(s-z_1)(s-z_2)\ldots(s-z_m)}{(s-p_1)(s-p_2)\ldots(s-p_n)} = \dfrac{2}{(s+1)(s+2)}$

State space form: $\begin{array}{l}\dot{\mathbf{x}} = \mathbf{A}\mathbf{x} + \mathbf{b}u \\ y = \mathbf{c}^{\mathrm{T}}\mathbf{x} + du\end{array}; \mathbf{A} = \begin{bmatrix} -3 & -1 \\ 2 & 0 \end{bmatrix}; \mathbf{b} = \begin{bmatrix} 1 \\ 0 \end{bmatrix}; \mathbf{c}^{\mathrm{T}} = [0 \quad 1]; d = 0$

Using the MATLAB commands tf, zpk and ss, the *LTI* system can be given in an *LTI* data-structure.

Defining the LTI sys structure

Let the transfer function of the system be $H(s) = \frac{2}{s^2+3s+2} = \frac{2}{(s+1)(s+2)}$.
The transfer function form is given as

```
num=2
den=[1, 3, 2]
H=tf(num,den)
```

The transfer function is:

```
         2
-------------
s^2 + 3 s + 2
```

or directly

```
Htf=tf(2,[1, 3, 2])
```

Defining the zero-pole-gain form:

```
Hzpk=zpk([],[-1, -2],2)
```

The zero/pole/gain form is:

```
     2
----------
(s+1)(s+2)
```

The state space form can be given by

```
A=[-3, -1; 2, 0]; B=[1; 0]; C=[0, 1]; D=0;
Hss=ss(A,B,C,D)
```

The models can be converted into each other:

```
H=zpk(H)
H=ss(H)
H=tf(H)
```

The *LTI* models possess several properties. These properties can be obtained using the command get.

```
get(Htf)
get(Hzpk)
get(Hss)
```

The *LTI sys* model parameters can be obtained by using the commands tfdata, zpkdata, ssdata. The *LTI* data structure can be used also in case of *Multi-Input Multi-Output* (*MIMO*) systems, therefore it stores some parameters in *cell array* form. The parameters can be accessed in vector format if we put the flag 'v' in the command.

```
[num,den]=tfdata(H,'v')
     num = 0      0     2
     den = 1      3     2
[z,p,k]=zpkdata(H,'v')
[A,B,C,D]=ssdata(H)        % here flag 'v' is omitted.
```

Symbolic data entry

The transfer function can be defined even more simply by symbolic data entry. Let us give the s variable of the LAPLACE transform by a special command:

```
s=zpk('s')
H=1/(s^2+3*s+2)
```

The transfer function appears in zero-pole-gain form.
If the variable s is given in the form

```
s=tf('s')
```

then the transfer functions defined with this variable will be obtained in tf form, i.e. in polynomial-polynomial form.

Arithmetic operations can be applied to data given in *LTI sys* structures as well. The most frequently used operations are: +, -, *, /, \, ', inv, ^. For example the resulting transfer function of a closed loop system can be calculated by the following symbolic relation:

```
Hcl=H/(1+H)
```

The possible simplifications are executed by the command minreal.

```
Hcl=minreal(Hcl
```

Among the *LTI sys* structures a hierarchical sequence order is defined: tf ->zpk ->ss. If in a command or in a calculation the operands are *LTI* models of different forms, then the result is always in the form which is higher in the hierarchy. For example, the result of

```
Htf*Hzpk
```

is obtained in the zpk form.

1.2.3 Time Domain Analysis

The *Control System Toolbox* contains several commands that provide basic tools for time domain analysis. Let us analyse the system

$$H(s) = \frac{2}{s^2+2s+4}$$

```
H=2/(s^2+2*s+4)
```

Define the time vector to be

```
t=0:0.1:10;
```

Step response: All previously discussed versions of the step command can be used. Additionally the following forms can also be applied:

```
step(H);
[y,t,x]=step(H);
```

Let us remark that when using the *LTI sys* structure the order of the output parameters differ from the order when using the (num, den) polynomial form:

```
[y,x,t]=step(num,den);
```

Impulse response: The impulse response is the response of the system to a DIRAC delta input.

```
impulse(H);
yi=impulse(H,t);
plot(t,yi)
```

The system's behaviour can also be analysed for nonzero *initial conditions*. Nonzero initial conditions can only be taken into account if state space models are used. Accordingly, to apply the initial command, the system has to be transformed into a state space representation.

```
H=ss(H)
x0=[1, -2]
[y,t,x]=initial(H,x0);
plot(t,y),grid on
```

Note that these commands yield x as a matrix having as many columns as dictated by the number of the state variables (two in this case), and as many rows as dictated by the time instants (109 in this case). Just to check:

```
size(x)
    ans = 109    2
```

The state trajectory can also be calculated and plotted. The first column of x contains the first state variable, while the second state variable will show up in the second column. The notation ':' means that all elements of a vector are chosen. The state trajectory plots a state variable versus the other one.

```
x1=x(:,1); x2=x(:,2);
plot(x1,x2)
```

Output response to an arbitrary input: The output can be calculated as a response for any input signal.

Let us determine the output signal if the input is the following sinusoidal signal: $u(t) = 2\sin(3t)$

```
usin=2*sin(3*t);
ysin=lsim(H,usin,t);
```

Plot the input and the output in the same diagram (input: red, output: blue).

```
plot(t,usin,'r',t,ysin,'b'), grid;
```

1.2.4 Frequency Domain Analysis

The system's behaviour can also be analysed in the frequency domain.

The BODE diagram can be calculated by the bode command. There are several ways to use this command. The gain and phase shift of the system can be calculated at a given frequency. Let us calculate the gain and the phase shift of system H at the frequency $\omega = 5$:

```
w=5;
[gain,phase]=bode(H,w);
```

The result is gain = 0.0860, phase = −154.5367.

These calculations can be repeated for several frequencies. The command bode can be activated to calculate the absolute value and the phase angle of the frequency function at several frequencies by one call. The BODE diagram can be displayed (see Fig. 1.4) by

```
bode(H),grid
```

In this case, MATLAB™ automatically calculates a frequency vector based on the system dynamics.

The frequency scale is logarithmic, as in this case a big frequency range can be taken into account. The calculations can be repeated for a selected frequency range. A logarithmic frequency vector can be generated by the logspace command

```
w=logspace(-1,1,200);
```

This command creates 200 logarithmically equidistant frequency points between $10^{-1} = 0.1$ and $10^{1} = 10$.

The values of the BODE diagram can be calculated at these frequency points as

```
[gain,phase]=bode(H,w);
```

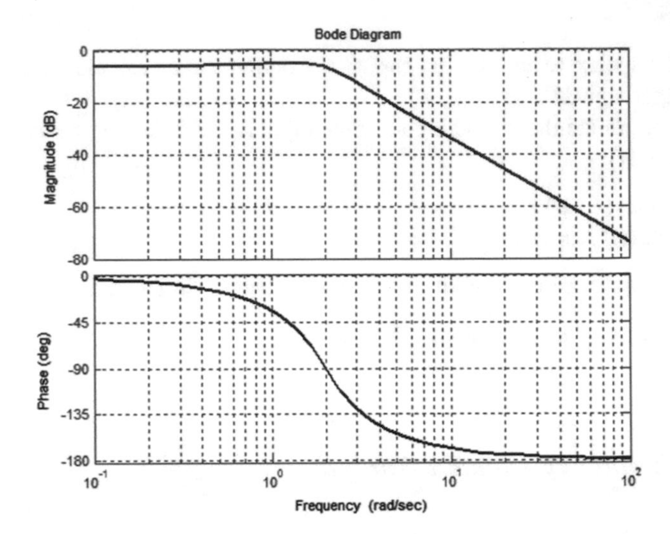

Fig. 1.4 BODE diagram

Since the *LTI* structure can be used also in the case of *MIMO* systems, the parameters *gain* and *phase* are given in 3 dimensional array format. This can be transformed to vector form by the operation (:). Let us compare the following two commands:

```
gain
gain(:)
```

NYQUIST diagram: At a given frequency the gain and the phase angle provide a vector (a point) in the complex plane. These vectors are plotted in the complex plane and their points are connected while the frequency is changing in a given range (Fig. 1.5).

The NYQUIST diagram is produced by the command

```
nyquist(H);
```

The command margin evaluates the main characteristics of the frequency function. It is an important tool to check the stability margins of a system.

```
margin(H);
```

(Frequency functions will be analyzed in more detail in Sect. 2.3.)

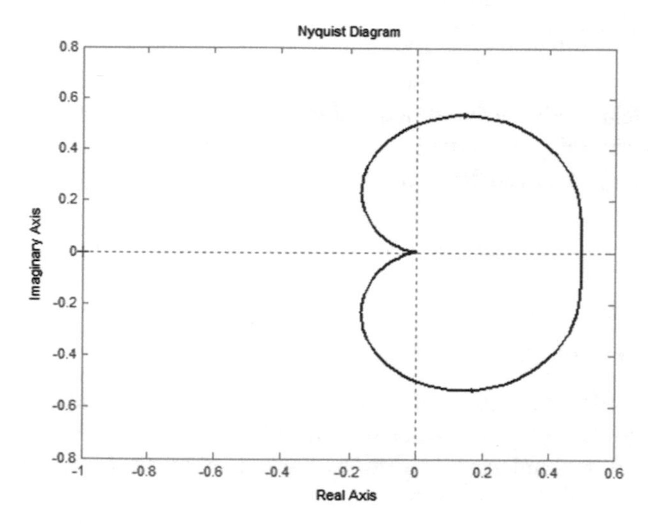

Fig. 1.5 Nyquist diagram

1.2.5 Zeros, Poles

The roots of the denominator of the transfer function are the poles of the system.

```
[num,den]=tfdata(H,'v');
poles=roots(den);
```

The roots of the numerator of the transfer function are the zeros of the system.

```
zeros=roots(num);
```

The zeros and the poles can be immediately obtained from the zpk model:

```
[z,p,k]=zpkdata(H,'v');
```

The zeros and the poles can be plotted in the complex plane:

```
subplot(111);
pzmap(H);
```

The damp command lists all the poles and (in the case of complex pole-pairs) the natural frequencies and damping factors:

```
damp(H);
```

The gain of the system (its steady state value in the case of a step input) can be calculated

```
K=dcgain(H);
```

1.2.6 LTI *Viewer*

A linear system can be analysed in detail by *LTI Viewer*, which is a graphical user interface for analysing the system response in the time domain and in the frequency domain. The systems can be analysed from the menu or using the right mouse button:

```
ltiview
```

or

```
ltiview('bode',H);
```

To demonstrate the application of *LTI Viewer*, we will first analyse a so called *first-order lag element* which can be described by a first-order differential equation. Its transfer function, differential equation, and step response can be obtained by the following relations.

First-order lag element:
The transfer function is the ratio of the LAPLACE transforms of the output and the input signals.

$$\frac{Y(s)}{U(s)} = H(s) = \frac{A}{1+sT}, \text{ or } (1+sT)Y(s) = AU(s).$$

Hence the differential equation is

$$y(t) + T\dot{y}(t) = Au(t)$$

and its solution for a unit step input is

$$y(t) = A\left(1 - e^{-t/T}\right)1(t).$$

The step response can be obtained in several ways using MATLAB™. Possibilities for simulation include the following:

a. by solving the differential equation:

```
T=10; A=5; t=0:0.1:50; y1=A*(1-exp(-t/T)); plot(t,y1);
grid;
```

b. on the basis of the transfer function:

```
y2=step(A,[T 1],t); plot(t,y1,t,y2)
```

c. using the *LTI* description:

```
s=tf('s'); P1=A/(1+T*s); y3=step(P1,t);
plot(t,y1,t,y2,t,y3)
```

d. using the block orientated SIMULINK™ program (see Sect. 1.3.).

It can be seen that the curves of the step responses calculated in the three different ways coincide.

With *LTI Viewer* a system can be imported and then analysed with its different characteristic functions. Let us consider the previous *LTI* model:

```
P1=A/(1+T*s)
```

The transfer function is

```
     5
--------
10 s + 1
```

```
ltiview
```

Select File/Import
Import from Workspace
Select P1
OK
RightClick, Select 'Plot Types':

Step
Impulse
Bode
Nyquist
Pole/Zero

...

Checking the points of the curves: Left Click on the curve
Analysing several systems in parallel:

```
P2=5/(1+20*s), P3=5/(1+50*s);
```

Back to the *LTI* viewer:
Import P1, P2, P3
Right Click, Select Systems: automatic order of the colours: blue, green, red
The system dynamics can be seen for the different time constants in the time- and the frequency domain and on the complex plane.

Let us now analyse the behaviour and characteristic functions of the so called second-order oscillating element, which can be described by a second-order differential equation.

```
P4=1/(s^2+s+1), P5=1/(s^2+0.5*s+1);ltiview
```

Let us add a zero to the second-order system and analyse its effect on the characteristic functions of the system.

```
s=zpk('s');
P6=8/6*(s+6)/(s+2)/(s+4),P7=8/3*(s+3)/(s+2)/(s+4);
ltiview
P8=8*(s+1)/(s+2)/(s+4);P9=-8/3*(s-3)/(s+2)/(s+4);
ltiview
```

1.3 SIMULINK™

SIMULINK™ is a graphics software package supporting block-oriented system analysis. SIMULINK™ has two phases, *model definition* and *model analysis*. First a model has to be defined, then it can be analysed by running a simulation. SIMULINK™ represents dynamical systems with block diagrams. Defining a system is much like drawing a block diagram. Instead of drawing the individual blocks, blocks are copied from libraries of blocks. The standard block library is organized into several subsystems, grouping blocks according to their behaviour. Blocks can be copied from these or any other libraries or models into your model. The SIMULINK™ block library can be opened from the MATLAB™ command window by entering the command simulink. This command displays a new window containing icons for the subsystem blocks. To construct your model, select **New** from the **File** menu of SIMULINK to open a new empty window in which you can build your model. Open one or more libraries and drag some blocks into your active window, then release the button. To connect two blocks use the left mouse button to click on either the output or input port of one block, drag to the other block's input or output port to draw a connecting line, and then release the button. By clicking on the block with the right button you can duplicate it. The blocks can be increased, decreased, and rotated. Open the blocks by double clicking to change some of their internal parameters. Save the system by selecting **Save** from the **File** menu. Figure 1.6. shows the SIMULINK™ diagram of a control system.

Run a simulation by selecting *Start* from the *Simulation* menu or by clicking on the *Run* icon (►). Simulation parameters can also be changed. You can monitor the behavior of your system with a *Scope* or you can use the *To Workspace* block to

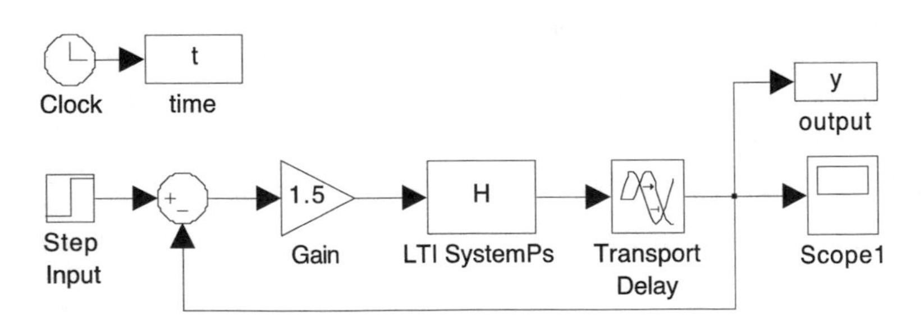

Fig. 1.6 SIMULINK™ diagram of a control system

send data to the MATLAB™ workspace and perform MATLAB™ functions (e.g. plot) on the results. Parameters of the blocks can be referred also by variables defined in MATLAB™. Simulation of SIMULINK™ models involves the numerical integration of sets of ordinary differential equations. SIMULINK™ provides a number of integration algorithms for the simulation of such equations. The appropriate choice of method and the careful selection of simulation parameters are important considerations for obtaining accurate results. To get yourself familiarized with the flavour of the options offered by SIMULINK™ consider the following example.

Create a new file and copy various blocks (Fig. 1.6). The block parameters should then be changed to the required value. Change the *Simulation –>Parameters–>Stop* time parameter to 50 from the menu. SIMULINK™ uses the variables defined in the MATLAB™ workspace.

$H(s)$: *Control System Toolbox –>LTI system* : H
Creating difference: *Simulink–>Math–>Sum*: +–
Dead-time, delay: *Simulink–>Continuous–>Transport Delay*: 1
Gain: *Simulink–>Math–>Gain*: 1.5
Step input: *Simulink–>Sources–>Step*
Scope: *Simulink–>Sinks–>Scope*
Clock: *Simulink–>Sources–>ClockOutput, time*: *Simulink–>Sinks–>To Workspace*: t, y

The result can be analysed directly by the *Scope* block or it can be sent back to the MATLAB™ workspace by the *To Workspace* output block. The results can be further processed and displayed graphically. Change the *Gain* parameter between 0.5 and 2. Determine the critical value of the gain, where steady oscillations do appear in the control system.

The results of the simulation can be sent to the MATLAB™ workspace through the *Scope* block as well. Let us set the parameters of the graphical window of the *Scope* as follows:

Under the '*properties*' menu

Data history: *Save data to workspace –>Variable name*: ty (tu for the control signal)

Matrix format

So the values of the time vector t and the output vector y can be obtained easily after the simulation, and then some properties (as e.g. the overshoot, settling time, maximum value of the control signal, etc.) can be determined.

```
t=ty(:,1)
y=ty(:,2)
plot(t,y),grid on
```

Chapter 2
Description of Continuous Systems in the Time-, Operator- and Frequency Domains

The behaviour of linear systems can be described in the time-, in the LAPLACE operator-, and in the frequency domain. The most straightforward information about the operation of practical systems is obtained by analysis in the time domain. The analysis in the frequency domain gives deeper insight into important properties of the systems. The design of control systems is frequently executed based on considerations in the frequency domain. In the LAPLACE operator domain the calculations related to the performance of the system become simpler than in the time domain. These domains can be converted to each other (Fig. 2.1).

2.1 Relationship Between the Time- and the Frequency Domain

A signal can be investigated in the frequency and in the time domain. Investigation in the frequency domain means that the signal is considered as a sum of sinusoidal components. Let us approximate a periodical rectangular signal by the sum of 4 sinusoidal signals. The odd coefficients of the frequency spectrum (FOURIER expansion) of the signal are: $4/\pi, 4/3\pi, 4/5\pi, 4/7\pi$. The approximation can be calculated by the MATLAB™ commands

```
w0=1; Ts=0.2;
t=0:Ts:51;
y=4/pi*(sin(w0*t)+ sin(3*w0*t)/3+ sin(5*w0*t)/5+
sin(7*w0*t)/7);
figure(1),plot(t,y);
```


L. Keviczky et al., *Control Engineering: MATLAB Exercises*, Advanced Textbooks in Control and Signal Processing,
https://doi.org/10.1007/978-981-10-8321-1_2

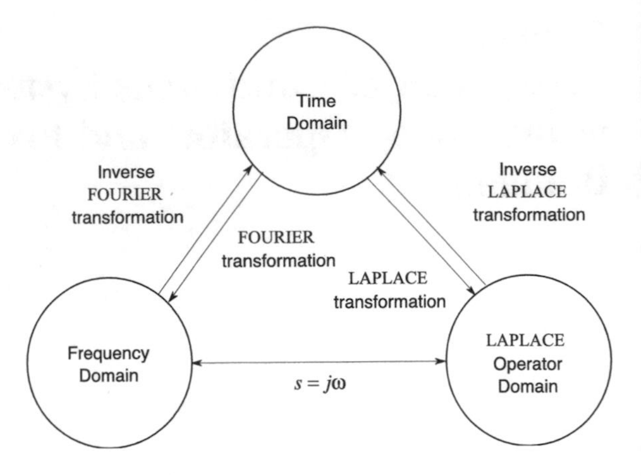

Fig. 2.1 Domains of calculations

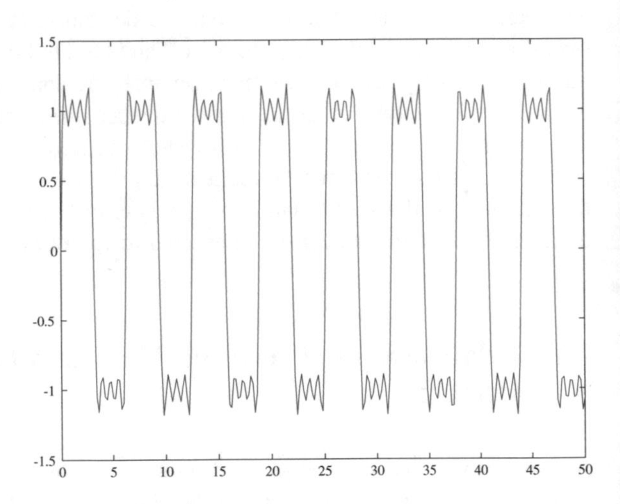

Fig. 2.2 A periodic signal approximated by 4 Fourier components

It can be seen that the approximation is already good with only 4 components (Fig. 2.2).

Let us plot the absolute value of the spectrum of the signal.

The command fft determines the Fourier transform of the signal.

```
Yf=fft(y);
n=length(t);nh=floor(n/2);
Yf=Yf(1:nh+1);
w=2*pi*(1:nh+1)/(n*Ts);
figure(2),plot(w,abs(Yf))
```

It can be seen that the spectrum (Fig. 2.3) contains values only at odd frequencies.

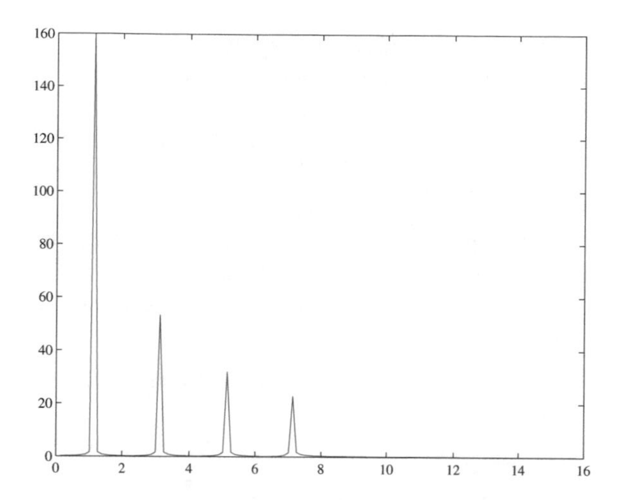

Fig. 2.3 Frequency spectrum

2.2 LAPLACE and Inverse LAPLACE Transformations

Analysing the behaviour of linear systems in the LAPLACE operator domain using LAPLACE transformation and the inverse LAPLACE transformation is easier than analysis in the time domain. LAPLACE transforms of the most frequently applied input signals:

$$\delta(t) \leftrightarrow 1 \quad ; \quad 1(t) \leftrightarrow 1/s \quad \text{and} \quad t \leftrightarrow 1/s^2.$$

Determine the step response of a system (Fig. 2.4).

In the LAPLACE operator domain the output signal is obtained by multiplication, $Y(s) = H(s)U(s)$, where $Y(s) = \mathcal{L}\{y(t)\}$ is the LAPLACE transform of the output signal $y(t)$ and $H(s)$ is the transfer function of the system, which is defined as the ratio of the LAPLACE transforms of the output and the input signal: $H(s) = \frac{Y(s)}{U(s)}$. The output signal is obtained by applying the inverse LAPLACE transformation: $y(t) = \mathcal{L}^{-1}\{Y(s)\}$.

Let us calculate the step response of the system given by the transfer function

$$H(s) = \frac{-2s^3 - 9s^2 - 5s + 18}{(s+2)(s+3)^2}.$$

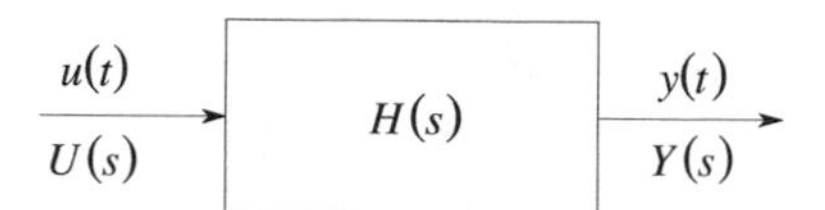

Fig. 2.4 System described by its transfer function

The input signal is a unit step, whose LAPLACE transform is $U(s) = \mathcal{L}\{1(t)\} = \frac{1}{s}$. The LAPLACE transform of the output signal is

$$Y(s) = U(s)H(s) = \frac{1}{s}\frac{-2s^3 - 9s^2 - 5s + 18}{(s+2)(s+3)^2}$$

The output signal $y(t)$ in the time domain can be obtained by inverse LAPLACE transformation. The LAPLACE transform of the signal is expanded to a sum of components whose LAPLACE transforms are known. The most common elements are

$$k \xrightarrow{\mathcal{L}^{-1}} k\,1(t) \quad ; \quad t \geq 0$$

$$\frac{r}{s+p} \xrightarrow{\mathcal{L}^{-1}} re^{-pt}$$

$$\frac{r}{(s+p)^2} \xrightarrow{\mathcal{L}^{-1}} rte^{-pt}$$

This form can be obtained by the partial fractional expansion of the LAPLACE transform of the output signal. In MATLAB™ this is executed by the command residue.

First give the LAPLACE transform of the output signal by the polynomials of its numerator and denominator.

```
s=zpk('s')
Y=(-2*s^3-9*s^2+-5*s+18)/(s*(s+2)*(s+3)*(s+3))
[num,den]=tfdata(Y,'v')
```

The polynomials can be given directly, as well.

```
num=[-2 -9 -5 18]
den=poly([-3 -3 -2 0])
```

Expansion in terms of partial fractions:

```
[r,p,k]=residue(num,den)
   r =  1.0000
        2.0000
       -4.0000
        1.0000
   p = -3.0000
       -3.0000
```

```
   -2.0000
    0
k = []
```

That means the result in the LAPLACE operator domain is

$$Y(s) = \frac{r(1)}{s-p(1)} + \frac{r(2)}{[s-p(2)]^2} + \frac{r(3)}{s-p(3)} + k = \frac{1}{s+3} + \frac{2}{(s+3)^2} - \frac{4}{s+2} + \frac{1}{s}$$

and in the time domain,

$$y(t) = e^{-3t} + 2te^{-3t} - 4e^{-2t} + 1(t), \quad t \geq 0$$

Let us observe the structure corresponding to the double pole in the vectors r and p in the partial fractional representation of the LAPLACE transform of the output signal and in the expression of the output signal in the time domain. The number of partial fractions belonging to a multiple pole is equal to the multiplicity of the pole.

Based on the analytical expression above the time function can be given in numerical form as

```
t=0:0.05:6;
y=r(1)*exp(p(1)*t)+r(2)*t.*exp(p(2)*t)+r(3)*exp(p(3)*t) +r(4)*exp(p(4)
*t);
```

In the second term on the right side the point besides t means that the operation is to be executed on the elements of the vector.

The values of $y(t)$ can be determined numerically even more simply.

```
yi=impulse(Y,t);
plot(t,y,t,yi),grid;
```

In the figure only one curve is seen, as the two curves coincide exactly.

Exercise:

Determine the inverse LAPLACE transform when there are conjugate complex poles.

$$Y(s) = \frac{2}{s^2 + 2s + 1.25}.$$

Find an analytical expression for the signal $y(t)$.

2.3 The Frequency Function

A basic property of a stable linear system is that for a sinusoidal input, it responds with a sinusoidal signal of the same frequency *in steady (quasi-stationary) state*. Applying the input signal

$$u(t) = A_u \sin(\omega t + \varphi_u) \quad t \geq 0,$$

the output signal is obtained as the sum of a quasi-stationary and a transient component.

$$y(t) = y_{steady}(t) + y_{transient}(t)$$

The output signal in quasi-stationary state (Fig. 2.5) is

$$y_{steady}(t) = A_y \sin(\omega t + \varphi_y)$$

The frequency function defines the *amplitude ratio* A_y/A_u and the *phase shift* $\varphi_y - \varphi_u$ as a function of frequency. Using the amplitude ratio and the phase shift within one single function the frequency function is derived as a complex function. It can be proven that formally the frequency function can be obtained from the transfer function by substituting $s = j\omega$.

$$H(j\omega) = H(s)|_{s=j\omega} = M(\omega)\, e^{j\varphi(\omega)}$$

$M(\omega)$ is the *amplitude function* (the absolute value of the frequency function) and $\varphi(\omega)$ is the *phase function*.

$$M(\omega) = |H(j\omega)| = \frac{A_y(\omega)}{A_u(\omega)} \quad ; \quad \varphi(\omega) = \arg\{H(j\omega)\} = \varphi_y(\omega) - \varphi_u(\omega)$$

The frequency function can be depicted in a given frequency range by plotting $M(\omega)$ and $\varphi(\omega)$ versus the frequency. The frequency scale is logarithmic. This technique gives the Bode diagram. A second possibility is to plot the points corresponding to pairs of $M(\omega)$ and $\varphi(\omega)$ of the frequency function calculated for various values of ω in the complex plane, while ω varies from zero to infinity. Connecting these points results in the contour of the so-called Nyquist diagram.

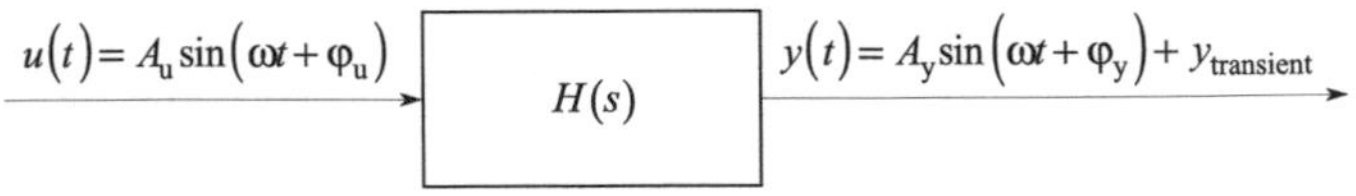

$u(t) = A_u \sin(\omega t + \varphi_u)$ → $H(s)$ → $y(t) = A_y \sin(\omega t + \varphi_y) + y_{transient}$

Fig. 2.5 System response to a sinusoidal input

2.3.1 *Calculation and Visualization of the Frequency Function*

Suppose the transfer function of a system is

$$H(s) = \frac{10}{s^2 + 2s + 10}.$$

Determine its output signal if the input signal is $u(t) = A_u \sin(\omega t)$, $A_u = 1, \ \omega = 3$.

```
num=10
den=[1, 2, 10]
H=tf(num,den)
t=0:0.05:10;
u=sin(3*t);
y=lsim(H,u,t);
```

Plot both the input (red) and output (blue) in the same diagram:

```
plot(t,u,'r',t,y,'b'), grid;
```

In steady, quasi-stationary state, after the decrease of the transient, the output signal is sinusoidal, its frequency is the same as that of the input signal, but its amplitude and phase angle differ from those of the input signal. Their values depend on the frequency. From the figure one sees ($M(3) = 1.64, \ \varphi = -80°$). The gain and the phase angle can be calculated from the frequency function $H(s = j\omega)$. In MATLAB™, the command bode can be employed to calculate these values at a given frequency or over a given frequency range. E.g. at $\omega = 3$,

```
[M,fi]=bode(H,3);
```

The values of the gain and the phase angle can be obtained from the complex frequency function as well.

$$H(j\omega) = \frac{10}{(j\omega)^2 + 2j\omega + 10} = \frac{10}{10 - \omega^2 + 2j\omega}$$

$$H(j3) = \frac{10}{10 - 3^2 + 2j3} = \frac{10}{1 + 6j}$$

```
H3=10/(1+6j)
M=abs(H3)
fi=angle(H3)*180/pi
```

Let us repeat the calculations if the frequency of the input signal is changed to $\omega = 10$. It can be seen that the values of the gain and the phase angle have changed.

The command bode plots the amplitude and the phase angle versus the frequency.

```
bode(H);
```

Check on the curve if at frequency $\omega = 3$ the gain and the phase angle are equal to the previously calculated values. The gain is given in decibels. The value of the gain $M(3) = 1.64$ in decibels is

```
20*log10(1.64)
```

The scale on the amplitude curve can be set from decibels to absolute values. Let us right click with the mouse on the white background of the amplitude diagram of the BODE diagram, then set on the appearing menu window *Properties, Units, magnitude in –> absolute*. Then the gain corresponding to the given frequency can be read directly from the amplitude diagram.

2.3.2 *Plotting the BODE and the NYQUIST Diagrams*

The bode command shows the amplitudes of the BODE diagram in decibels and the phase angles in degrees. The frequency scale is logarithmic.

```
bode(H)
```

Let us calculate the amplitude and the phase values in vector format and then plot the diagram.

```
[gain,phase,w]=bode(H)
```

The bode command determines automatically the points of the frequency vector based on the poles and zeros of the system. These values are provided in the vector w on the left side of the command. If we would like to calculate the frequency function over a different frequency range, the frequency vector can be given by the command logspace which determines a row vector with logarithmically equidistant frequency points.

```
w=logspace(-1,2,200)
```

The first two parameters of logspace give the lower and the upper points of the frequency range in powers of 10. The above command calculates 200 logarithmically equidistant points between the lower point $10^{-1} = 0.1$ and the upper point $10^2 = 100$ (without giving the third variable, the command employees 50 points). If, e.g. the upper point of the frequency range is 300, the command is called in the following form:

```
w=logspace(-1,log10(300),200)
```

Let us repeat the calculation of the Bode diagram with this frequency vector.

```
[gain,phase]=bode(H,w)
```

(Remark: the variables on the right side of a MATLAB™ command are the input variables, while the variables on the left side are the output variables.)

The *LTI sys* structure generates three dimensional matrices (because of the possible *MIMO* systems). With the (:) operator the results can be transformed to vector form.

```
gain=gain(:),phase=phase(:)
```

The amplitude and the phase diagrams can be plotted in different windows of the screen by the command subplot. (E.g. subplot(211) generates 2×1 windows on the screen and refers to the first one.)

Plot the amplitude and the phase angle in linear scale, and the frequency with logarithmic scale by calling the command semilogx. This command is used similarly to plot.

```
subplot(211),semilogx(w,gain)
subplot(212),semilogx(w,phase)
```

Generally the amplitude is plotted in logarithmic scale, using the command loglog.

```
subplot(211),loglog(w,gain)
subplot(212),semilogx(w,phase)
```

On the amplitude scale, the powers of 10 do appear. To convert the values to decibels use the following commands.

```
subplot(211),semilogx(w,20*log10(gain)),grid
subplot(212),semilogx(w,phase),grid
```

The BODE diagram is advantageous when multiplying two transfer functions (calculating the resulting transfer functions of serially connected elements). Because of the logarithmic scale the BODE diagrams are just added. In most cases, approximate BODE diagrams, given by the asymptotes of the magnitude curve, provide a good approximation of the frequency characteristics. By sketching these approximate curves, a quick evaluation of the system's behaviour can be made.

The NYQUIST diagram plots the points of the frequency function in the complex plane. Its shape characterizes the system. The important properties of the system can be determined by analysing it.

```
nyquist(H)
```

Calling the command without variables on the left side plots the NYQUIST diagram extending the curve with points calculated for negative frequencies. This is the so called entire or total NYQUIST diagram.

The real and the imaginary components can be calculated from the values of the amplitudes and the phase angles.

```
re=real(gain.*exp(j*phase*pi/180));
im=imag(gain.*exp(j*phase*pi/180));
```

The real and imaginary values belonging to the different frequency values can also be calculated directly, with the command nyquist.

```
[re,im]=nyquist(H,w);
re=re(:);im=im(:);
```

Then the Nyquist diagram for positive frequencies can be plotted.

```
plot(re,im)
```

To supplement the curve with the part belonging to negative frequencies, it has to be throwing back to the real axis.

```
re2=[re;flipud(re)]
im2=[im;flipud(-im)]
plot(re2,im2)
```

2.4 Operations with Basic Elements

In a control system, the basic connections of elements are the series connection, parallel connection, and feedback. With block diagram algebra a complex system can be built with these basic connections.

Given the following two systems with their transfer functions:

$$H_1(s) = \frac{10s+1}{s+1} \quad \text{and} \quad H_2(s) = \frac{s}{(10s+1)(5s+1)}$$

Determine the resulting transfer functions

- of the serially connected systems,
- of the parallel connected systems,
- when H_1 is fed back through a unity gain element (negative feedback)
- when H_1 is fed back through H_2 (negative feedback)

Let us define two systems in MATLAB™.

```
s=zpk('s')
H1=(10*s+1)/(s+1)
H2=s/((10*s+1)*(5*s+1))
```

Serially connected systems (Fig. 2.6):

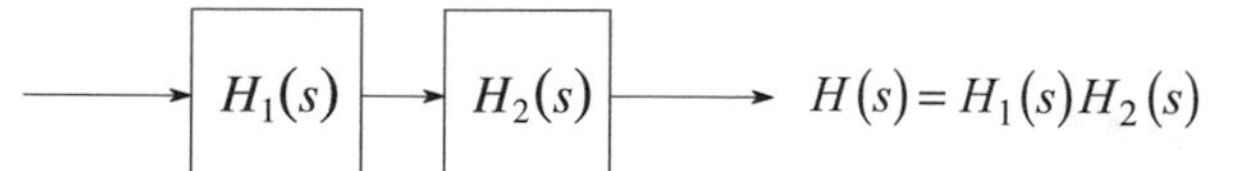

Fig. 2.6 Serially connected systems

The resulting transfer function:

```
H=H1*H2
```

The series command also calculates the resulting transfer function of a series connection:

```
H=series(H1,H2)
        0.2 s (s+0.1)
     ---------------------
     (s+1) (s+0.2) (s+0.1)
```

It can be seen that the numerator and the denominator have common roots (zeros, poles) which can be cancelled using the command minreal.

```
H=minreal(H)
          0.2 s
     -------------
     (s+1) (s+0.2)
```

Parallel connected systems (Fig. 2.7):
The resulting transfer function:

```
H=H1+H2
     10 (s+0.06876) (s^2 + 0.3332 s + 0.02909)
     -----------------------------------------
             (s+1) (s+0.2) (s+0.1)
```

Negative feedback through unit gain (Fig. 2.8):

Fig. 2.7 Parallel connection

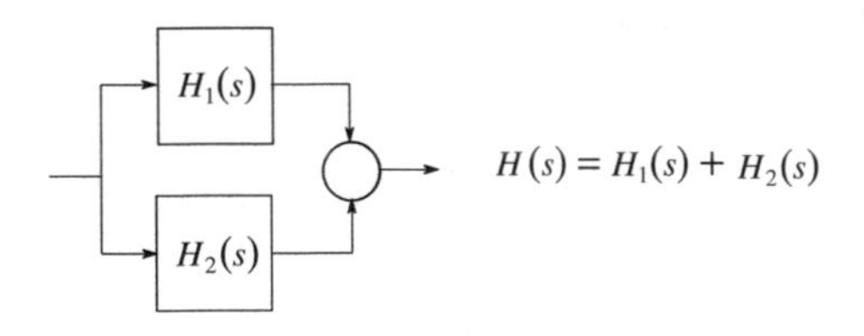

Fig. 2.8 Negative feedback through unit gain

$H_1(s)$

$$H(s) = \frac{H_1(s)}{1+H_1(s)}$$

The resulting transfer function:

```
H=H1/(1+H1)
```

The command feedback can also be applied to calculate the resulting transfer function. The second parameter is the transfer function in the feedback path, the third parameter shows that the feedback is negative.

```
H=feedback(H1,1,-1)
(or H=feedback(H1,1), the basic definition is negative feedback.)
H=minreal(H)
        0.90909 (s+0.1)
        ---------------
          (s+0.1818)
```

Negative feedback (Fig. 2.9):
The resulting transfer function:

```
H=H1/(1+H1*H2)
```

With the command feedback:

```
H=feedback(H1,H2,-1)
H=minreal(H)

    10 (s+0.1) (s+0.2)
    ---------------------
    (s+1.239) (s+0.1615)
```

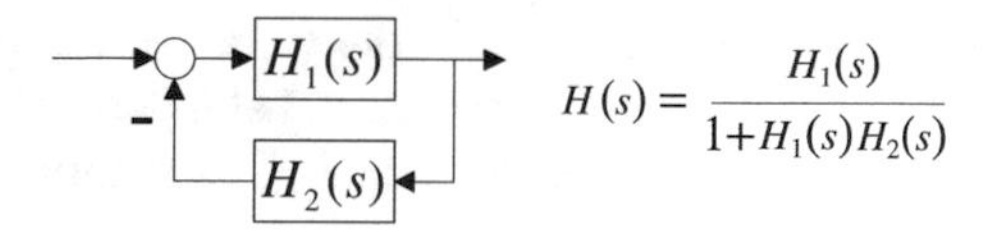

Fig. 2.9 Negative feedback

2.5 Basic Elements of a Linear System

A linear system generally can be given in the following time constant form:

$$H(s) = \frac{K}{s^i} \frac{\prod_1^c (1 + s\tau_j) \prod_1^d \left(1 + 2\zeta_j \tau_{oj} s + s^2 \tau_{oj}^2\right)}{\prod_1^e (1 + sT_j) \prod_1^f \left(1 + 2\xi_j T_{oj} s + s^2 T_{oj}^2\right)} e^{-sT_d}$$

where K is the gain, i is the number of the integrators, T_d is the dead-time, τ_o and T_o are time constants, and ζ and ξ are the damping factors.

In the sequel the time- and frequency characteristics of the most important elements will be analysed. These are the proportional, integrating, differentiating, dead-time, and lag elements, and more complex elements obtained by series connections of these basic elements.

2.5.1 *Proportional (P) Element*

$$H(s) = H_P(s) = K$$

The gain is K, the phase angle is zero at all frequencies.

2.5.2 *Integrating (I) Element*

$$H(s) = H_I(s) = \frac{K}{s}$$

Here $i = 1$: the system contains an integrator. The integrator has a "memory": its output value depends on the values of the past inputs. Its output can be constant only if the input value is zero. Let us investigate the properties of a pure integrator given by a transfer function $H_I(s)$ for gains $K = 1$ and $K = 5$.

$$H_1(s) = \frac{1}{s} \quad \text{and} \quad H_2(s) = \frac{5}{s}.$$

```
clear            % clear all the previously defined variables.
s=zpk('s')       % define the symbolic s LAPLACE variable in zpk form.
H1=1/s
H2=5/s
```

The step response:

```
figure(1), step(H1,'r',H2,'g'), grid
```

BODE diagram:

```
figure(2), bode(H1,'r',H2,'g'), grid
```

NYQUIST diagram:

```
figure(3), nyquist (H1,'r',H2,'g'),grid
```

It can be seen that the step response increases linearly. The amplitude of the frequency function at low frequencies is infinity. Its phase angle is $-90°$ for all frequencies.

2.5.3 *First-Order Lag Element (PT1)*

$$H(s) = H_{\mathrm{T}}(s) = \frac{K}{1 + Ts}$$

Determine the step response and the BODE and NYQUIST diagrams of the *PT1* element given by the transfer function

$$H(s) = \frac{2}{1 + 10s}.$$

Define the system by

```
H=2/(1+10*s)
```

or

```
H=tf(2,[10, 1])
```

The step response:

```
t=0:0.1:50;
y=step(H,t);
plot(t,y),grid
```

or simply

```
step(H)
```

Let us investigate the effect of the parameters K and T on the system response.

The steady value of the output signal, $y(t \to \infty)$, can be calculated by the final value theorem of the LAPLACE transformation:

$$y(t \to \infty) = \lim_{s \to 0} sY(s) = \lim_{s \to 0} sH(s)U(s),$$

where $U(s)$ is the LAPLACE transform of the input signal. For unit step input,

$$y(t \to \infty) = \lim_{s \to 0} s\, H(s)U(s) = \lim_{s \to 0} s\, H(s)\frac{1}{s} = \lim_{s \to 0} H(s)$$

In the case of the considered system,

$$y(t \to \infty) = \left.\frac{2}{1+10s}\right|_{s=0} = 2.$$

With MATLAB™:

```
yinf=dcgain(H)
```

Investigate the effect of the time constant parameter T on the transient behaviour. Repeat the command

```
y=step(2,[T 1],t);
```

for several different values of T.

The transfer function of the system can be given in *zero-pole form* as well:

$$H_1(s) = \frac{k_p}{s-p} = \frac{2}{10(0.1+s)} = \frac{0.2}{s+0.1},$$

where

$$p = -\frac{1}{T} \quad \text{and} \quad k_p = \frac{k}{T}$$

The absolute value of the pole gives the break-point frequency of the approximating BODE amplitude-frequency diagram.

The steady-state value of the step response

$$y(t \to \infty) = \lim_{s \to 0} H(s)$$

coincides with the low frequency value of the BODE amplitude-frequency function.

The BODE and NYQUIST diagrams of the system are obtained by the following commands:

```
bode(H);
nyquist(H);
```

The characteristic functions of the following first-order, second-order and third-order (*PT1*, *PT2*, *PT3*) elements can be calculated similarly.

$$H_1 = \frac{2}{1+10s}; \quad H_2 = \frac{2}{(1+10s)(1+2s)}; \quad H_3 = \frac{2}{(1+10s)(1+2s)(1+s)}$$

(1—is in red, 2—is in green, 3—is in blue)

```
H1=2/(1+10*s)
H2=2/((1+10*s)*(1+2*s))
H3=2/((1+10*s)*(1+2*s)*(1+s))
```

Step responses:

```
figure(1), step(H1,'r',H2,'g',H3,'b'),grid
```

BODE diagrams:

```
figure(2), bode(H1,'r',H2,'g',H3,'b'), grid
```

NYQUIST diagrams:

```
figure(3), nyquist(H1,'r',H2,'g',H3,'b'),grid
```

With more lags the step response is slower.

Remark: in the figure window the marked part of the plots can be enlarged by command *zoom*.

As it was shown previously, the characteristic functions of several elements can be investigated simultaneously also by the LTI Viewer.

2.5.4 Second-Order Oscillating (ξ) Element

$$H(s) = H_\xi(s) = \frac{1}{s^2T_o^2 + 2\xi T_o s + 1}$$

Let us investigate the system:

$$H(s) = \frac{1}{9s^2 + 2s + 1} = \frac{1}{s^2T_o^2 + 2\xi T_o s + 1}$$

where $\omega_o = \frac{1}{T_o}$ is the natural frequency and ξ is the damping factor ($T_o = \frac{1}{\omega_o} = 3$, $\xi = 1/3$).

```
num=1;
den=[9, 2, 1]
H=tf(num,den)
```

The poles of the system are calculated by the command roots,

```
roots(den)
```

or by the command damp:

```
damp(H)
```

The conjugate complex poles can be given in the following form: $p_1 = a + jb$, $p_2 = a - jb$.

The overshoot v_t of the step response is calculated by

$$\omega_o^2 = a^2 + b^2 \quad ; \quad \xi = -\frac{a}{\omega_o} \quad \text{and} \quad v_t = e^{\frac{-\xi\pi}{\sqrt{1-\xi^2}}} = e^{-\frac{\pi a}{b}}.$$

The oscillation frequency is

$$\omega_p = b = \omega_o\sqrt{1-\xi^2}$$

The time of the first maximum of the step response (the peak time) is $T_p = \pi/\omega_p = \pi/b$.

```
kszi=1/3
vt=exp(-kszi*pi/sqrt(1-kszi*kszi))
```

The step response can be obtained as follows:

```
[y,t]=step(H);
plot(t,y), grid
```

The maximum value of the step response:

```
ym=max(y)
```

Its steady state value:

```
ys=dcgain(H)
```

The overshoot can be calculated also as

```
yo=(ym-ys)/ys
```

Let us analyse the step responses, the Bode diagram and the Nyquist diagram, for several values of the damping factor: $\xi = 0.3, 0.7, 1, 2$.

```
kszi1=0.3, kszi2=0.7, kszi3=1, kszi4=2
T0=3
H1=1/(s*s*T0*T0+2*kszi1*T0*s+1)
H2=1/(s*s*T0*T0+2*kszi2*T0*s+1)
H3=1/(s*s*T0*T0+2*kszi3*T0*s+1)
H4=1/(s*s*T0*T0+2*kszi4*T0*s+1)
```

The step responses:

```
figure(1), step(H1,'r',H2,'g',H3,'b',H4,'m'),grid
```

The Bode diagrams:

```
figure(2), bode(H1,'r',H2,'g',H3,'b',H4,'m'),grid
```

The Nyquist diagrams:

```
figure(3), nyquist(H1,'r',H2,'g',H3,'b',H4,'m'),grid
```

The pole-zero configurations:

```
figure(4), pzmap (H1,'r',H2,'g',H3,'b',H4,'m')
```

The poles can be obtained also with command damp:

```
damp(H1)
damp(H2)
damp(H3)
damp(H4)
```

It can be seen that for damping factor $\xi = 0.3$ the step response is the most oscillating, the maximum amplification in the Bode amplitude diagram is the highest, and the Nyquist diagram crossing the imaginary axis gives the biggest magnitude for this case. The poles are complex conjugates. The imaginary value of the complex conjugate poles providing the frequency of oscillation in the time response is also the highest. High amplification in the Bode amplitude diagram indicates a high overshoot in the step response. If this should be avoided, no high amplification is allowed in the Bode amplitude diagram. The damping factor $\xi = 0.7$ provides a slight overshoot. Control systems can be designed for similar behaviour. For $\xi = 1$ the system has two coinciding real poles. In the case of $\xi > 1$ there are two different real poles, and the step response is aperiodic. There is no overshoot in the step response and no amplification in the Bode amplitude diagram.

2.5.5 *Differentiating (D and DT) Elements*

The transfer function of the ideal differentiating element is $H(s) = sT_{\mathrm{d}}$.

```
H=s
bode(H)
step(H)
  ??? Error using ==> rfinputs
  Not supported for non-proper models.
```

MATLAB™ can not evaluate the system responses as the element is non realizable, its transfer function is non-proper, the degree of its numerator is higher than that of its denominator. Its step response is the Dirac delta.

The differentiating effect can be realized only together with lag elements.

$$H_1(s) = \frac{2s}{1+10s}; \quad H_2(s) = \frac{2s}{(1+10s)(1+2s)}$$

Give the step responses, the Bode and the Nyquist diagrams of these elements.

```
H1=(2*s)/(1+10*s)
H2=(2*s)/((1+10*s)*(1+2*s))
```

Step responses:

```
figure(1); step(H1,'r',H2,'b'),grid
```

Bode diagrams:

```
figure(2), bode(H1,'r',H2, 'b'),grid
```

Nyquist diagrams:

```
figure(3), nyquist(H1,'r',H2, 'b'),grid
```

It can be seen that a differentiating element behaves as a high pass filter. It supresses the DC (low frequency) component of a signal and amplifies the high frequency components.

2.5.6 *The Effect of Zeros*

Suppose the transfer function is given in the following form. The roots of the numerator are the zeros of the transfer function.

$$H(s) = \frac{k(s - z_1)(s - z_2)\ldots(s - z_m)}{\mathrm{D}(s)}$$

Let us analyse how the zeros affect the step response and the frequency response in case of the following transfer function:

$$H(s) = \frac{1 + \tau s}{(1 + s)(1 + 10s)}$$

The time constant τ in the numerator (the zero is $-1/\tau$) changes between -12 and 12. In the case of a positive zero (which is located in the right half-plane of the complex plane) the system is called a non-minimum phase system.

```
s=tf('s')
D=(1+s)*(1+10*s)
tau=[-12 -4 0 4 12]
for i=1:5,H(i)=(s*tau(i)+1)/D,end
figure(1),step(H(1),'r',H(2),'g',H(3),'k',H(4),'m',H(5),'b')
```

The step responses are seen in Fig. 2.10. Note that in the case of a non-minimum phase system, they behave unexpectedly. E.g. in the case of one right-side zero the step response starts in the opposite direction related to the steady state value, then it

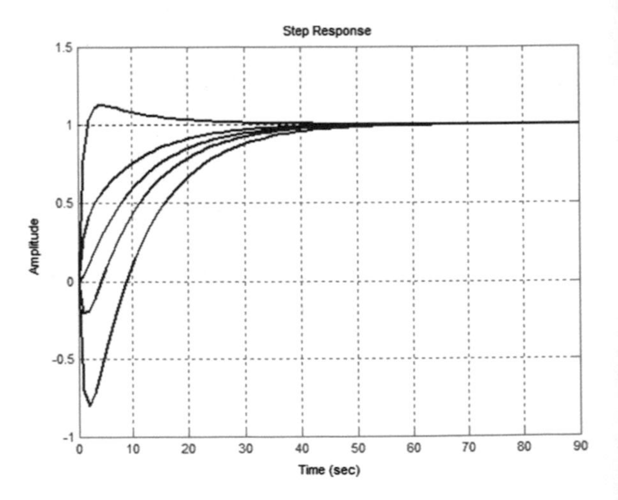

Fig. 2.10 Zeros affect the step response

changes direction and reaches the steady state value. Inserting a zero in the system results in an accelerated time response.

Bode and Nyquist diagrams:

```
figure(2),bode(H(1),'r',H(2),'g',H(3),'k',H(4),'m',H(5),'b)
figure(3),nyquist(H(1),'r',H(2),'g',H(3),'k',H(4),'m',H(5),'b)
```

Let us evaluate the effect of a zero in the frequency domain, how it influences the Bode and the Nyquist diagrams.

2.5.7 Dead-Time Element

Its transfer function is

$$H_{\mathrm{H}}(s) = H(s)e^{-sT_{\mathrm{d}}}$$

Its description in the time and in the Laplace operator domain is

$$y(t) \to y(t - T_{\mathrm{d}}); \quad Y(s) \to Y(s)e^{-sT_{\mathrm{d}}}$$

In the frequency domain, this is

$$|e^{-j\omega T_{\mathrm{d}}}| = 1, \quad \arg\{e^{-j\omega T_{\mathrm{d}}}\} = -\omega T_{\mathrm{d}} \text{ (in radians)}.$$

Its gain is calculated as $|H_{\mathrm{H}}| = |H|$, and its phase angle is $\arg(H_{\mathrm{H}}) = \arg(H) - \omega T_{\mathrm{d}}$.

Let us analyse the frequency functions of the elements

$$H_1(s) = \frac{1}{1+10s} \quad \text{and} \quad H_2(s) = \frac{1}{(1+10s)}e^{-2s}.$$

The amplitude and the phase angle of the dead-time element are

$$|H_1| = |H_2| \quad , \quad \text{gain2} = \text{gain1}$$

$$\arg(H_2) = \arg(H_1) - \omega T_{\mathrm{d}} \quad , \quad \text{phase2} = \text{phase1} - \text{w*Td}$$

```
Td=2
H1=1/(1+10*s)
num1=1
den1=[10, 1]
```

Now when calculating the Bode diagram, use the polynomial form given by the num and den numerator and denominator polynomials.

First let us generate the frequency vector.

```
w=logspace(-2,2,500);
[gain1,phase1]=bode(H1,w);
```

Change the gain1, phase1 values to vector form.

```
gain1=gain1(:);phase1=phase1(:);
phasedelay=180/pi*Td*w';
```

The bode command calculates the phase angle in degrees. The phase delay $\varphi = -\omega T_d$ of the dead-time element is obtained in radians. Therefore it has to be converted to degrees.

The amplitude and the phase angle considering the dead-time can be obtained as follows:

```
gain2=gain1;
phase2=phase1-phasedelay;
subplot(211),loglog(w, gain1,'r',w, gain2,'b'),grid;
subplot(212),semilogx(w, phase1,'r',w, phase2,'b'),grid
```

The linearity of the course of the phase angle can be seen better if drawing it command plot is used instead of semilogx.

```
figure(2),subplot(111),plot(w,phase1,'r',w,phase2,'b'),grid
```

Now let us plot the Nyquist diagram. First calculate the real and imaginary values of the frequency function.

```
h1= gain1.*exp(j*phase1*pi/180);
h2= gain2.*exp(j*phase2*pi/180);
figure(2),plot(real(h1),imag(h1),'r',real(h2),imag(h2),'b')
```

The behaviour in the high frequency domain could be seen better if a bigger frequency range is given with the command logspace.

The behaviour of a dead-time element in the time domain can be investigated better in SIMULINK™, as the time-delay is offered as a single building block.

The transfer function of the dead-time element is not a rational function. Nevertheless it can be approximated by a non-minimum phase rational fraction where the first elements of its TAYLOR expansion are the same as those of the exponential transfer function characterizing the dead-time. These rational functions are called PADE functions. The higher the degree of the PADE function, the better is the approximation. It has to be mentioned that with this approximation the step response starts with +1 or –1 instead of zero. In MATLAB™, the command pade calculates the approximation.

Demonstrating the use of PADE approximation, use a 5-th order approximation.

$$H_{\mathrm{H}}(s) = e^{-sT_{\mathrm{d}}} \cong H_{\mathrm{PADE}}(s)$$

```
pade(Td,5);
```

As there is no output parameter, now this command shows graphically the step response.

```
[numd,dend]=pade(Td,5)
Hd=tf(numd,dend)
Hd=zpk(Hd)
H2=H1*Hd
```

The step responses:

```
figure(1), step(H1,'r',H2,'g'),grid
```

The BODE diagram:

```
figure(2), bode(H1,'r',H2,'g'),grid
```

The NYQUIST diagram:

```
figure(3), nyquist(H1,'r',H2,'g')
```

It should be emphasized that in the frequency domain it is better to consider the phase modifying effect of the dead-time than to employ the PADE approximation. In the time domain the analysis can be executed better by running the simulation in SIMULINK™.

Problem: Let us investigate how good is the PADE approximation in the above case of a first-order lag element with dead-time. Build a SIMULINK™ program using the "*transport delay*" block, and using the fifth-order PADE rational function. Compare the step responses.

2.5.8 Evaluation of the Characteristics of the Elements, the Effects of Poles and Zeros

The values of the step responses in steady state ($t \to \infty$) and the values of the amplitude response of the frequency function for $\omega \to 0$ are the same.

NYQUIST diagrams of proportional elements at $\omega = 0$ start from a point of the positive real axis, which characterizes the gain of the element. The NYQUIST diagram of an integrating element at $\omega = 0$ starts from infinity in the direction of the negative imaginary axis. The NYQUIST diagram of a double integrating element starts from infinity in the direction of the negative real axis. NYQUIST diagrams of derivative elements start from the zero point of the complex plane in the direction of the positive imaginary axis. In case of a transfer function containing only lags (no zeros), the NYQUIST diagram covers as many quarters in the complex plane as there are time lags. The zeros deteriorate the monotonic change of the phase angle. The BODE amplitude diagram of a proportional element starts parallel to the frequency axis with zero phase angle, the BODE amplitude diagram of a system containing one integrator starts with a slope of –20 dB/decade with a -90° phase angle, whereas the BODE diagram of a system with two integrators starts with a slope of −40 dB/decade and -180° phase angle. Time lags break down the slope of the BODE amplitude diagram by −20 dB/decade, while zeros make the slope go up by +20 dB/decade.

A time constant is the reciprocal of the pole. If the pole is located far away from the origin at the left side of the real axis, this means a fast transient behaviour. In the frequency domain it affects the frequency function at higher frequencies.

Chapter 3
State-Space Representation of Continuous Systems

In the time domain, linear systems can be characterized by their input and output signals. Far more information is obtained, however, if there are also considered those (mostly internal) signals whose value remains unchanged in case a step-like change is applied in the input signal. These signals represent information determined by the history of the development of the system and they do not exhibit sudden changes in their character. The set of these signals are called state variables (or states for short) and the models employing states are called state-space models (*SSM*). Using an *SSM*, the set $\{\boldsymbol{A}, \boldsymbol{b}, \boldsymbol{c}^{\mathrm{T}}, d\}$ exhibits an *SISO* system representation with input u, output y and state vector $\boldsymbol{x}$ in the model

$$\dot{\boldsymbol{x}} = \boldsymbol{A}\boldsymbol{x} + \boldsymbol{b}u$$
$$y = \boldsymbol{c}^{\mathrm{T}}\boldsymbol{x} + du$$

where the matrices $\{\boldsymbol{A}, \boldsymbol{b}, \boldsymbol{c}^{\mathrm{T}}, d\}$ are parameter matrices describing the system. Formally, *SSM* describe linear systems with n first-order differential equations and the *SSM* consists of a set of these first-order differential equations as one single first order vector differential equation and an output equation. The state equation is a differential equation (so it needs to be solved), while the output equation is only a linear combination using the states and the input signal (Note that in the state-space model $\boldsymbol{A}$ is an $n \times n$ quadratic matrix, $\boldsymbol{b}$ is an $n \times 1$ column vector, $\boldsymbol{c}^{\mathrm{T}}$ is a $1 \times n$ row vector, and d is an 1×1 dimension constant.).

3.1 State Transformation

It is well known that there are infinitely many of input-output equivalent *SSM* associated with a given transfer function. MATLAB™ offers one possible way to get a *SSM*, using the command `tf2ss` ('transfer function to state space').

L. Keviczky et al., *Control Engineering: MATLAB Exercises*, Advanced Textbooks in Control and Signal Processing,
https://doi.org/10.1007/978-981-10-8321-1_3

Example 3.1.1 Start our discussion with the transfer function

$$H(s) = \frac{1}{s^2 + 3s + 2} = \frac{1}{(s+2)(s+1)}$$

and transform it to an *SSM*:

```
s = zpk('s')
H = 1/(s*s + 3*s + 2)
H = ss(H)
        a =
                  x1  x2
             x1   -1   1
             x2    0  -2
        b =
                  u1
             x1    0
             x2    1
        c =
               x1  x2
          y1   1   0

        d =
                 u1
             y1  0
```

The step response can be obtained by

```
step(H);
```

The parameter matrices can also be retrieved from the *LTI* form:

```
[A,b,c,d] = ssdata(H)
```

and the step response can also be obtained from the parameter matrices:

```
step(A,b,c,d);
```

Further on, the parameter matrices can be retrieved from the polynomial form, as well:

```
num = 1,den = [1 3 2]
[A1,b1,c1,d1] = tf2ss(num,den)
    A1 =

          -3   -2
           1    0
    b1 =

           1
           0
    c1 =

           0    1
    d1 =

           0
```

It can be seen that the two state space representations are different from each other. Still they give the same input-output transfer function. Just to check this, derive the transfer functions from the state models:

```
Hchk = ss(A,b,c,d);Hchk = zpk(Hchk)
Hchk1 = ss(A1,b1,c1,d1);Hchk1 = zpk(Hchk1)
```

The polynomial form of the transfer function can be obtained from the state space form as

```
[num,den] = ss2tf(A1,b1,c1,d1)
```

From a given *SSM*, another different representation can be generated using a coordinate transformation in the state space. The state variables in the new coordinate-system can be obtained by a linear transformation (called a *similarity transformation*). In more detail, assuming $\boldsymbol{T}$ to be a non-singular quadratic transformation matrix and z the new state variable of the transformed system, the similarity transformation can be summarized in the following set of equations:

$$\dot{z} = \tilde{\boldsymbol{A}}z + \tilde{\boldsymbol{b}}u$$
$$y = \tilde{\boldsymbol{c}}^{\mathrm{T}}z + \tilde{d}u$$

where

$$z = \boldsymbol{Tx} \quad \Rightarrow \quad \boldsymbol{x} = \boldsymbol{T}^{-1}z$$
$$\tilde{\boldsymbol{A}} = \boldsymbol{TAT}^{-1}, \tilde{\boldsymbol{b}} = \boldsymbol{Tb}, \tilde{\boldsymbol{c}}^{\mathrm{T}} = \boldsymbol{c}^{\mathrm{T}}\boldsymbol{T}^{-1}, \tilde{d} = d$$

Note, that if $\boldsymbol{T}$ is constructed so that the column vectors of $\boldsymbol{T}^{-1}$ are the eigenvectors of $\boldsymbol{A}$, the resulting matrix $\tilde{\boldsymbol{A}}$ will be a diagonal. In this case, the transformation is called a canonical transformation and the transformed state model is said to be in parallel canonical form.

Example 3.1.2 Find the canonical form of the system introduced in Example 3.1.1.

To properly build up the transformation matrix first determine the eigenvectors and deposit them in a V matrix:

```
[V,ev] = eig(A1)
   V =
       -0.8944   0.7071
        0.4472  -0.7071

   ev =
       -2      0
        0     -1
```

Then apply the relations derived for the similarity transformation:

```
Ti = V; T = inv(V)
Ap = T*A1*Ti
bp = T*b1
cp = c1*Ti
dp = d1
```

It can be seen that Ap is a diagonal, as expected.
The above steps can be replaced by the following more compact comands:

```
[Ap,bp,cp,dp] = ss2ss(A1,b1,c1,d1,inv(V))
```

or

```
[Ap,bp,cp,dp] = canon(A1,b1,c1,d1,'modal')
    Ap =
          -2      0
           0     -1
    bp =
          -2.2361
          -1.4142
    cp =
           0.4472  -0.7071
    dp =
            0
```

3.2 Solution of the State Equation by Analytical Methods

3.2.1 Solution of the State Equation in the Time Domain

$$\boldsymbol{x}(t) = e^{At}\boldsymbol{x}(0) + \int_0^t e^{A(t-\tau)}\boldsymbol{b}\,u(\tau)\mathrm{d}\tau = e^{At}\boldsymbol{x}(0) + \boldsymbol{x}_u(t)$$

where $\boldsymbol{x}_o(t)$ is the response to the initial conditions and $\boldsymbol{x}_u(t)$ the response to the input signal.

As e^{At} is defined by its TAYLOR series by

$$e^{At} = \boldsymbol{I} + \boldsymbol{A}t + \frac{1}{2}(\boldsymbol{A}t)^2 + \ldots + \frac{1}{n!}(\boldsymbol{A}t)^n + \ldots$$

it can be obtained in closed analytical form if $\boldsymbol{A}$ is a diagonal matrix. In this case, for typical input signals, like a unit step, the system response can be easily calculated. So it is worthwhile to transform both the system equations and the initial conditions to canonical form, then to derive the solution in this form, and finally to transform the results back into the original coordinate system.

Example 3.2.1 Let the initial conditions for the system introduced in Example 3.1.1 be given by $x_1(0) = 1$; $x_2(0) = 2$, or in vector form $\mathbf{x}(0) = \begin{bmatrix} 1 \\ 2 \end{bmatrix} = \mathrm{x0}$. Assume an input as a unit step: $u(t) = 1(t)$. Find $\mathbf{x}(t)$ for $t > 0$ in analytical form. Using a canonical transformation, the initial conditions for the state variables can be obtained as

```
x0 = [1 2]'
z0 = T*x0
```

where the transformation matrix calculated in Example 3.1.2 has been applied

```
T =
    -2.2361   -2.2361
    -1.4142   -2.8284
```

The transformed initial conditions are

```
z0 =
    -6.7082
    -7.0711
```

Then for the first component of the state vector of the canonical form we have:

$$z_o(t) = e^{A_p \cdot t}\begin{bmatrix} -6.7082 \\ -7.0711 \end{bmatrix} = e^{\begin{bmatrix} -2t & 0 \\ 0 & -t \end{bmatrix}}\begin{bmatrix} -6.7082 \\ -7.0711 \end{bmatrix} = \begin{bmatrix} e^{-2t} & 0 \\ 0 & e^{-t} \end{bmatrix}\begin{bmatrix} -6.7082 \\ -7.0711 \end{bmatrix} = \begin{bmatrix} -6.7082e^{-2t} \\ -7.0711e^{-t} \end{bmatrix}$$

while the first component of the state vector of the original system turns out to be:

$$\boldsymbol{x}_o(t) = \boldsymbol{T}^{-1}z(t) = \begin{bmatrix} -0.8944 & 0.7071 \\ 0.4472 & -0.7071 \end{bmatrix}\begin{bmatrix} -6.7082e^{-2t} \\ -7.0711e^{-t} \end{bmatrix} = \begin{bmatrix} 6e^{-2t} - 5e^{-t} \\ -3e^{-2t} + 5e^{-t} \end{bmatrix}.$$

The canonical state response for the unit step input becomes:

$$z_o(t) = \int_0^t e^{A_p(t-\tau)}\, \boldsymbol{bp}\, u(\tau)\mathrm{d}\tau = \int_0^t \begin{bmatrix} e^{-2(t-\tau)} & 0 \\ 0 & e^{-(t-\tau)} \end{bmatrix}\begin{bmatrix} -2.2361 \\ -1.4142 \end{bmatrix} 1(\tau)\mathrm{d}\tau$$

which gives

$$z_o(t) = \begin{bmatrix} -2.2361 \int_0^t e^{-2(t-\tau)} d\tau \\ -1.4142 \int_0^t e^{-(t-\tau)} d\tau \end{bmatrix} = \begin{bmatrix} -2.2361e^{-2t} \int_0^t e^{2\tau} d\tau \\ -1.4142e^{-t} \int_0^t e^{\tau} d\tau \end{bmatrix} = \begin{bmatrix} -1.11805(1 - e^{-2t}) \\ -1.4142(1 - e^{-t}) \end{bmatrix}$$

Transforming back to the original state coordinates yields

$$\begin{aligned} \boldsymbol{x}(t) = T^{-1}z(t) &= \begin{bmatrix} -0.8944 & 0.7071 \\ 0.4472 & -0.7071 \end{bmatrix}\begin{bmatrix} -1.11805(1 - e^{-2t}) \\ -1.4142(1 - e^{-t}) \end{bmatrix} \\ &= \begin{bmatrix} 1 - e^{-2t} - 1 + e^{-t} \\ -0.5(1 - e^{-2t}) + 1 - e^{-t} \end{bmatrix} \end{aligned}$$

which leads to

$$\boldsymbol{x}(t) = \begin{bmatrix} e^{-2t} + e^{-t} \\ 0.5 + 0.5e^{-2t} - e^{-t} \end{bmatrix}$$

The overall state response then is the sum of the response to the initial conditions and the response to the input signal. The output signal of the system is simply determined by $y = \boldsymbol{c}^{\mathrm{T}}\boldsymbol{x}$. Note the output can be calculated from the original or from the transformed state variables.

Example 3.2.2 For the problem discussed in Example 3.1.1, find the value of the state variables at $t = 5$ s. The initial values of the state variables are $x_1(0) = 1$ and $x_2(0) = 2$, or in vector form, $\mathbf{x}(0) = \begin{bmatrix} 1 \\ 2 \end{bmatrix} = \mathrm{x0}$. The input signal is zero. Using the MATLAB™ commands

```
A = [-3 -2;1 0]
t = 5;
x0 = [1 2]'
x5 = expm(A*t)*x0
```

results in

```
x5 =
   -0.0334
    0.0336
```

The state response for the initial conditions can also be calculated using the initial command:

```
b = [1;0]; c = [0 1];d = 0;
Hv = ss(A,b,c,d)
[y,t,x] = initial(Hv,x0)
plot(t,x(:,1),t,x(:,2)),grid
```

Beyond the state variables as function of time they can be plotted as state trajectories, e.g. x_1 as a function of x_2:

```
plot(x(:,1),x(:,2)),grid
```

A state trajectory allows us to think about the dynamic behaviour of the system.

3.2.2 Solution of the State Equation in the Laplace Operator Domain

The Laplace transform of the state vector is given by

$$\boldsymbol{x}(s) = (s\boldsymbol{I} - \boldsymbol{A})^{-1}\boldsymbol{x}(0) + (s\boldsymbol{I} - \boldsymbol{A})^{-1}\boldsymbol{b}\,U(s)$$

while the output signal is

$$Y(s) = \left[\boldsymbol{c}^{\mathrm{T}}(s\boldsymbol{I} - \boldsymbol{A})^{-1}\boldsymbol{b} + d\right]U(s).$$

The transfer function then turns out to be

$$P(s) = \frac{Y(s)}{U(s)} = \boldsymbol{c}^{\mathrm{T}}(s\boldsymbol{I} - \boldsymbol{A})^{-1}\boldsymbol{b} + d$$

Remark No matter which state representation form of the system we start from, the transfer function is the same.

Example 3.2.3 Find the transfer function of the system introduced in Example 3.1.1.

The parameter matrices of the system are

```
A =

    -3   -2
     1    0
b =

     1
     0
c =

     0    1
d =

     0
```

Calculating the transfer function results in

$$P(s) = \boldsymbol{c}^{\mathrm{T}}(s\boldsymbol{I} - \boldsymbol{A})^{-1}\boldsymbol{b} + d = \begin{bmatrix} 0 & 1 \end{bmatrix} \begin{bmatrix} s+3 & 2 \\ -1 & s \end{bmatrix}^{-1} \begin{bmatrix} 1 \\ 0 \end{bmatrix} + 0 = \frac{1}{s^2 + 3s + 2}.$$

MATLAB™ simply follows the analytical expression:

```
I = [1 0;0 1]
H = c1*inv(s*I-A1)*b1 + d1
```

or simply employes the ss2tf command:

```
[num,den] = ss2tf(A,b,c,d)
```

Problem Show that the same result is obtained if the canonical forms are used.

Problem Solve *Example* 3.2.1 in the LAPLACE operator domain.

3.3 Controllability, Observability

Controllability (more precisely, state controllability) is a notion to answer the question whether all the states can or can not be controlled by the input signal arbitrarily and independently from each other. The output controllability gives an answer to the question if the output signal can be controlled by the input signal arbitrarily or not. Observability tells if the initial value of the state variables can or can not be determined based on an input-output record of a certain time period.

3.3.1 Determination of Controllability and Observability Assuming Canonical Form

Controllability and observability can be determined directly if the canonical forms are available. An *SISO* linear system is (state) controllable if in its diagonal parameter matrix Ap all the values along the main diagonal (the eigenvalues) are different from each other and all the values in the vector **bp** are different from zero. An *SISO* linear system is observable if in its diagonal parameter matrix Ap all the values along the main diagonal (the eigenvalues) are different from each other and all the values in the vector **cp** are different from zero.

Example 3.3.1 Define the system parameter matrices as follows:

```
A = [-1 -0.5 0.5; 2 -3 0; 2 -1 -2]
b = [2;3;1]
c = [0 0 1]
d = 0
H = ss(A,b,c,d)
```

Find the canonical form of the system and check its controllability and observability.

To get the canonical form, first find the eigenvectors of the parameter matrix A and collect them into a matrix V:

```
[V,ev] = eig(A)
```

where eV gives the eigenvalues.

A canonical transformation applies the inverse of V as a transformation matrix:

```
Ti = V; T = inv(V)
Ap = T*A*Ti
bp = T*b
cp = c*Ti
dp = d
```

Note that Ap is diagonal, as was expected.

The above results can also be obtained using more compact MATLAB™ commands:

```
[Ap,bp,cp,dp] = ss2ss(A,b,c,d,inv(V))
```

or

```
[Ap,bp,cp,dp] = canon(A,b,c,d,'modal')
```

As a result the canonical representation becomes:

$$\begin{bmatrix} \dot{x}_1 \\ \dot{x}_2 \\ \dot{x}_3 \end{bmatrix} = \begin{bmatrix} -1 & 0 & 0 \\ 0 & -3 & 0 \\ 0 & 0 & -2 \end{bmatrix} \begin{bmatrix} x_1 \\ x_2 \\ x_3 \end{bmatrix} + \begin{bmatrix} 1.73 \\ 0 \\ -2.23 \end{bmatrix} u$$

$$y = \begin{bmatrix} 0.57 & -0.707 & 0 \end{bmatrix} \begin{bmatrix} x_1 \\ x_2 \\ x_3 \end{bmatrix} + 0u$$

Draw the block diagram of the system (Fig. 3.1). This is a *parallel representation* of the system. Notice that the gain is obtained by $b_1 c_1 = 1.73 \cdot 0.57 = 1$. It is

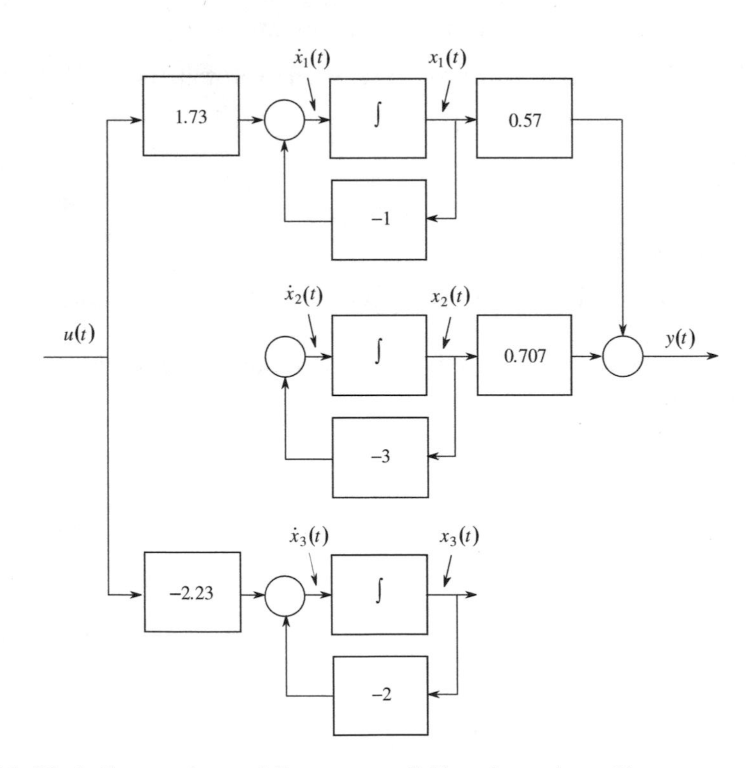

Fig. 3.1 Block diagram of a not fully state controllable and non-observable system

clear that any b_1 and c_1 satisfying $b_1c_1 = 1$ gives identical I/O equivalent state space representations.

The above parallel structure allows us to directly read the observability and controllability conditions. Namely x_3 is not observable (there is no information on x_3 in y) and x_2 is not controllable. The output signal is controllable as it is influenced by the input signal u through the state x_1.

3.3.2 Determination of Controllability and Observability from Arbitrary (Non-canonical) Representations

Controllability and observability are analytically handled by checking the rank of the appropriate KALMAN controllability and observability matrices (see Sect. 3.4 of the textbook [1]).

The controllability matrix is built up as follows:

$$\boldsymbol{M}_c = \begin{bmatrix} \boldsymbol{b} & \boldsymbol{A}\boldsymbol{b} & \dots & \boldsymbol{A}^{n-1}\boldsymbol{b} \end{bmatrix}.$$

A state-space representation is state controllable if the rank of the above matrix is n. Also, a state-space representation is output controllable if the row vector

$$\boldsymbol{m}_c^{\mathrm{T}} = \begin{bmatrix} \boldsymbol{c}^{\mathrm{T}}\boldsymbol{b} & \boldsymbol{c}^{\mathrm{T}}\boldsymbol{A}\boldsymbol{b} & \dots & \boldsymbol{c}^{\mathrm{T}}\boldsymbol{A}^{n-1}\boldsymbol{b} \end{bmatrix} = \boldsymbol{c}^{\mathrm{T}}\boldsymbol{M}_c$$

has at least one nonzero element (the rank of this matrix is equal to the number of the output signals).

The observability matrix is built up as follows: $\boldsymbol{M}_o = \begin{bmatrix} \boldsymbol{c}^{\mathrm{T}} \\ \boldsymbol{c}^{\mathrm{T}}\boldsymbol{A} \\ \vdots \\ \boldsymbol{c}^{\mathrm{T}}\boldsymbol{A}^{n-1} \end{bmatrix}.$

A state-space representation is observable if the rank of the above matrix is n.

Example 3.3.2 Check the controllability and observability of the system introduced in Example 3.3.1. Note that this is a third order state-space representation ($n = 3$).

The controllability matrix can be obtained by MATLAB™ using command ctrb as

```
Mc = ctrb(A,b)
```

or using the *LTI* structure:

```
Mc = ctrb(H)
```

The system is state controllable if the rank of $\boldsymbol{M}_c$ turns out to be $n = 3$.

```
rank(Co)
```

It is seen that this representation is not controllable, as $\text{rank}(\boldsymbol{M}_c) = 2 < n = 3$.

To check the output controllability use the following command:

```
rank(c*Mc)
```

As this value is 1, which is equal to the number of outputs (1), the system is output controllable.

To check the observability with MATLAB™ use the command obsv:

```
Mo = obsv(A,c)
Mo = obsv(H)
rank(Mo)
```

As $\text{rank}(\boldsymbol{M}_o) = 2 < n = 3$, the system is not observable.

Find the transfer function of the system.

```
[num,den] = ss2tf(A,b,c,d)
```

or in zero-pole form:

```
Hzpk = zpk(H)
[z,p,k] = zpkdata(H,'v')
```

Retrieve the zero-pole-gain information from the zero-pole form:

```
z =
    -3.0000
    -2.0000
p =
    -3.0000
    -2.0000
    -1.0000
k =
     1.0000
```

Note the double pole-zero cancellation in the transfer function.

```
Hzpk = minreal(Hzpk)
```

Using transfer function representation information gets lost, because the transfer function only reflects information on the controllable and observable subsystem of the complete state-space representation.

The residue command allows us to get the parallel transfer function form:

```
[num,den] = tfdata (H,'v')
[r,p,k] = residue(num,den)
```

$$H(s) = \frac{0}{s+3} + \frac{0}{s+2} + \frac{1}{s+1}$$

Note that the transfer function $\frac{1}{s+1}$ only partly describes the system represented by the state-space equations. This subsystem, is the (state) controllable and observable part of the system.

Example 3.3.3 Consider the following third order system:

$$H(s) = \frac{8}{(s+2)^3}.$$

The number of the states will be 3, however, as the system has repeated poles, these poles are not independent from each other, it is expected that the state model will not be controllable and will not be observable.

```
num = 8;
den = [1 6 12 8];
[A,b,c,d] = tf2ss(num,den)
[V,ev] = eig(A)
   V =
        0.8729      -0.8729      -0.8729
       -0.4364       0.4364       0.4364
        0.2182      -0.2182      -0.2182
   eV =
       -2.0000       0            0
            0       -2.0000       0
            0        0           -2.0000
```

It is seen the V eigenvectors are not linearly independent, consequently no canonical form can be obtained using similarity transformation. The following block diagram (Fig. 3.2) of the system helps us to come up with a close-to-parallel state-space representation.

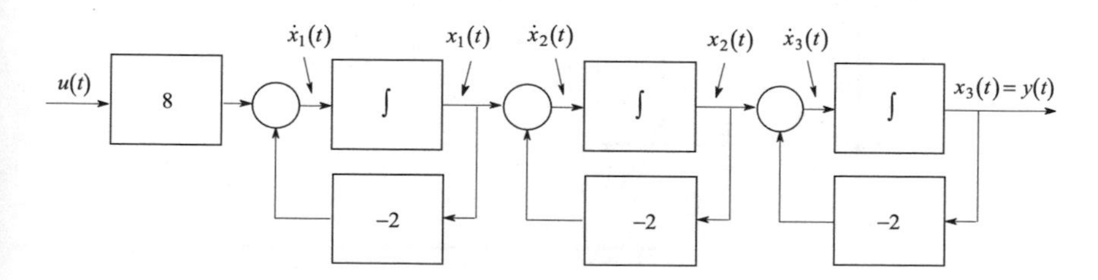

Fig. 3.2 State block diagram of a system with multiple poles

The related state equation:

$$\begin{bmatrix} \dot{x}_1 \\ \dot{x}_2 \\ \dot{x}_3 \end{bmatrix} = \begin{bmatrix} -2 & 0 & 0 \\ 1 & -2 & 0 \\ 0 & 1 & -2 \end{bmatrix} \begin{bmatrix} x_1 \\ x_2 \\ x_3 \end{bmatrix} + \begin{bmatrix} 8 \\ 0 \\ 0 \end{bmatrix} u$$

$$y = \begin{bmatrix} 0 & 0 & 1 \end{bmatrix} \begin{bmatrix} x_1 \\ x_2 \\ x_3 \end{bmatrix} + 0\,u$$

It is seen that in matrix $\boldsymbol{A}$ the poles show up along the main diagonal, however additional unity values accompany the pure diagonal form. This form is called JORDAN form.

Example 3.3.4 Consider the following closed-loop system (Fig. 3.3).

The process to be controlled is a first-order lag given by the transfer function:

$$H_2(s) = \frac{5}{10s+1} = \frac{0.5}{s+0.1}$$

This first-order lag can be equivalently redrawn as an integrator with a constant feedback (Fig. 3.4). Define the state variables as the output of the integrators in the complete system. The state equations can then be easily derived as

$$\begin{bmatrix} \dot{x}_1 \\ \dot{x}_2 \end{bmatrix} = \begin{bmatrix} -5.1 & 0.5 \\ -1 & 0 \end{bmatrix} \begin{bmatrix} x_1 \\ x_2 \end{bmatrix} + \begin{bmatrix} 5 \\ 1 \end{bmatrix} u$$

$$y = \begin{bmatrix} 1 & 0 \end{bmatrix} \begin{bmatrix} x_1 \\ x_2 \end{bmatrix} + 0\,u$$

Check the controllability and observability of the above state-space representation of the closed-loop system.

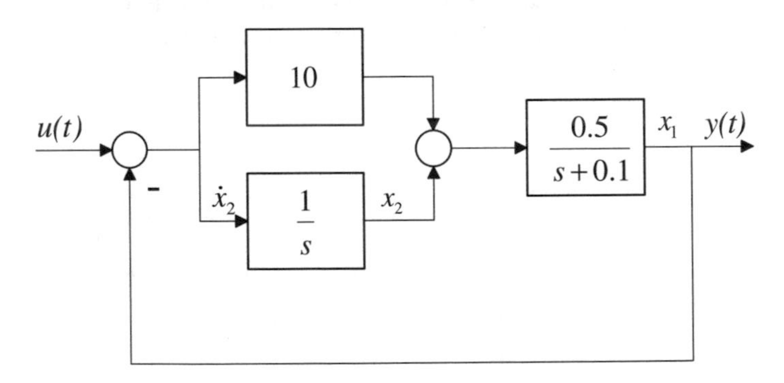

Fig. 3.3 Block diagram of a control system

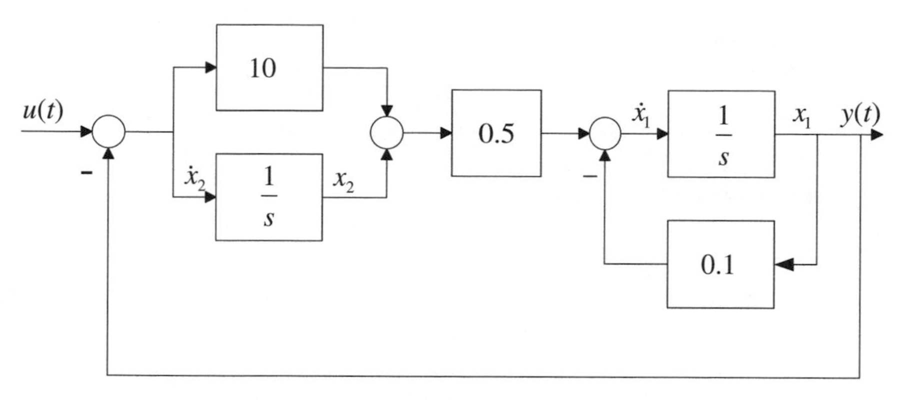

Fig. 3.4 Block diagram of the control system showing the state variables

The parameter matrices in MATLAB™ are given as

```
A = [-5.1 0.5;-1 0]
b = [5; 1]
c = [1 0]
d = 0
```

The rank of the controllability and observability matrices is calculated as

```
rank(ctrb(A,b))
    ans = 1
```

The system is not (state) controllable,

```
rank(c*ctrb(A,b))
    ans = 1
```

but the system is output controllable.

```
rank(obsv(A,c))
    ans = 2
```

The system is observable.

The reason why the system is not controllable can be seen by analysing the dynamics of the regulator, namely the zero of the regulator cancels the pole of the process, so one state variable becomes "invisible".

```
H = ss(A,b,c,d)
H = zpk(H)
      5 (s + 0.1)
     ----------------
     (s + 5) (s + 0.1)
H = minreal(H)
      5
     -----
     (s + 5)
```

Chapter 4
Negative Feedback

Feedback is the most important structure in control systems. The regulator gets information about the value of the controlled variable through feedback. Feedback significantly modifies the performance of the system.

4.1 Quality Characteristics and the Properties of Negative Feedback

The performance of a system controlled by negative feedback can be characterised numerically by its quality characteristics.

4.1.1 Requirements Set for Control Systems

Generally the following requirements are set for closed-loop control systems.

Stability: Stable operation of the control system is a basic requirement. Stability can be formulated in several ways. *Bounded Input–Bounded Output* (*BIBO*) stability means that the system provides a bounded output as a response to any and all bounded inputs. The system is asymptotically stable if its transients decay.

Robustness: The performance of a closed-loop system should not be sensitive to the inaccuracy of the available information about the process. Stability has to be guaranteed even if the parameters of the system are not known accurately or their values change within a given range around their nominal values.

L. Keviczky et al., *Control Engineering: MATLAB Exercises*,
Advanced Textbooks in Control and Signal Processing,
https://doi.org/10.1007/978-981-10-8321-1_4

Static behaviour: Another important requirement is the static accuracy of the system, i.e. its accuracy in steady state. Static requirements set for the steady state of the control system can include:

- Reference signal tracking. The tracking error should be below the prescribed value.
- Disturbance rejection: In steady state the control system should eliminate the effect of disturbances.

Static accuracy depends on the structure of the system and also on the input signals.

Transient response: The transient response is an important dynamic feature of a control system. The characteristic properties of the transient response can be given in the time domain by the characteristics of the transient response of the system in the case of a step reference signal or disturbance. The prescriptions for the transient response are the following:

- Overshoot of the output signal.
- Settling time: During the settling time the controlled variable approximates its steady value within 1–2%. Generally a small overshoot within 5–10% of the steady state value, can be tolerated, but there are processes where aperiodic transients are required.

Error integrals: In more complex control problems restrictions can be prescribed for the whole course of the output and the control signal. E.g. the quadratic integral values of the error signal should be minimised and the value of the control signal restricted.

Example 4.1 Analyse the effect of negative feedback and determine the quality properties of a closed-loop control circuit. The transfer function of the open-loop is

$$L(s) = \frac{4}{(10s+1)(4s+1)} = \frac{num}{den}$$

Unity negative feedback is applied (Fig. 4.1). The resulting transfer function of the closed loop is:

$$T(s) = \frac{L}{1+L} = \frac{num}{num+den}$$

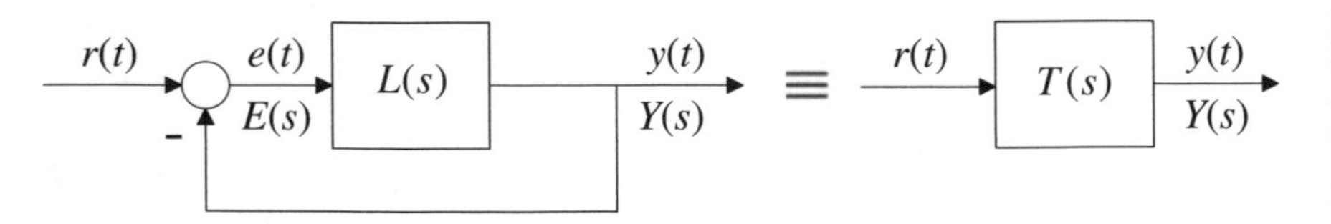

Fig. 4.1 Negative feedback

```
s = zpk('s')
L = 4/((10*s + 1)*(5*s + 1))
T = L/(1 + L)
T = minreal(T)
```

The command minreal is used to cancel the common poles and zeros.

The resulting (overall) transfer function can also be calculated by the command feedback. The second input parameter gives the transfer function in the feedback path.

```
T = feedback(L,1)
```

Let us compare the step responses of the open- and of the closed-loop (Fig. 4.2).

```
step(L,'b',T,'r'),grid on
```

It is evident that the closed-loop behaviour differs from that of the open-loop. The static and transient properties of the system were influenced significantly by the feedback. Let us determine the quality properties of the feedback system. The static value ($t \to \infty$) can be read from the figure. For the open-loop this value is 4, while in case of the closed-loop this is a bit less than 1. These values can be calculated more accurately by

```
ysL = dcgain(L)
ysT = dcgain(T)
    ysL = 4
    ysT = 0.8
```

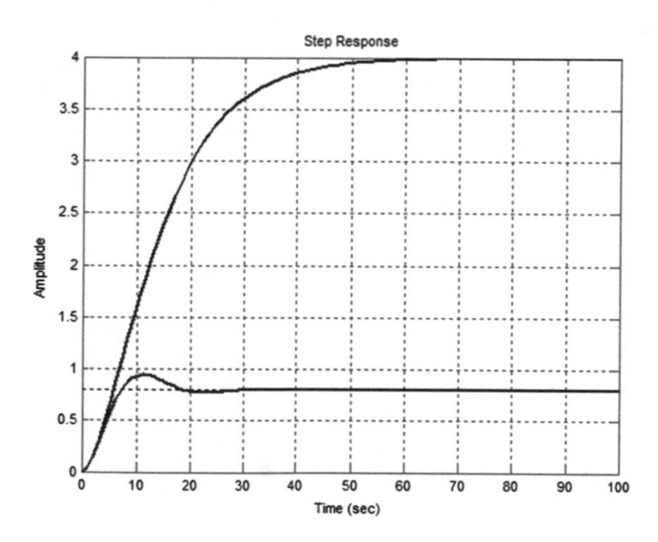

Fig. 4.2 Open- and closed-loop step responses

In control systems, reference signal tracking is one of the most important system characteristics. It is seen that in the case of negative feedback, a closed-loop system in steady state approximates the value of the reference signal. Let us calculate the steady state error.

```
esL = 1-ysL
esT = 1-ysT
    esL = -3
    esT = 0.2
```

In the figure it is also seen that the transient behaviour also has changed. The settling process became faster (the settling time can be read from Fig. 4.2).

$$t_{\text{sL}} \cong 45; \quad t_{\text{sT}} \cong 20.$$

There is an overshoot in closed-loop response, which can be calculated by

```
y = step(T)
ym = max(y)
yt = (ym-ysT)/ysT
```

The value of the overshoot is 18%.

Plot the Bode diagrams of the open-loop and the closed-loop in one diagram (Fig. 4.3).

```
bode(L,'b',T,'r'),grid on
```

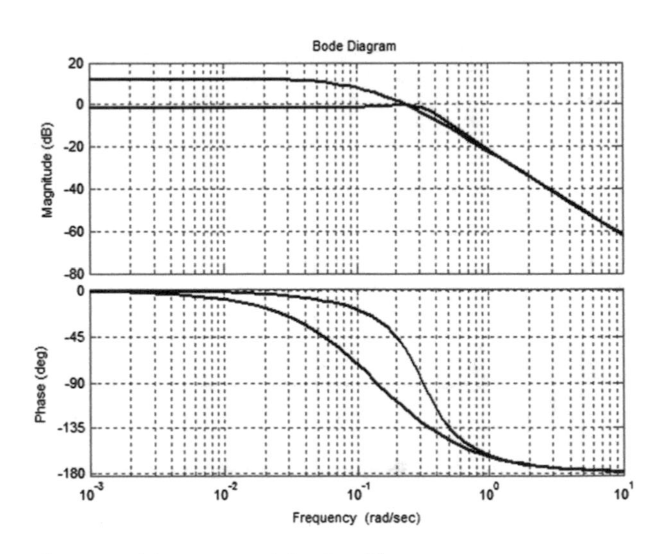

Fig. 4.3 Bode diagrams of the open- and the closed-loop

It can be seen that the BODE amplitude diagram of the closed-loop is approximately 1 in the low frequency domain, while in the high frequency domain it coincides approximately with the diagram of the open-loop.

Example 4.2 The transfer function of the closed-loop is

$$T(s) = \frac{1}{(1+10s)(1+s)}$$

Determine the linear error area from its step response.

The system is given by

```
T = 1/((10*s + 1)*(s + 1))
```

Plot the sampled points of its step response (Fig. 4.4).

The distance between two consecutive points is the sampling time Ts.

```
Ts = 0.5;
t = 0:Ts:60;
y = step(T,t);
plot(t,y,'.');grid on
```

The linear error area can be calculated by evaluating the integral

$$I_1 = \lim_{t \to \infty} \int_0^t e(\tau)\,\mathrm{d}\tau,$$

which can be calculated by using the relation

$$I_1 = A\left(\sum_{j=1}^{n} T_j - \sum_{k=1}^{m} \tau_k\right).$$

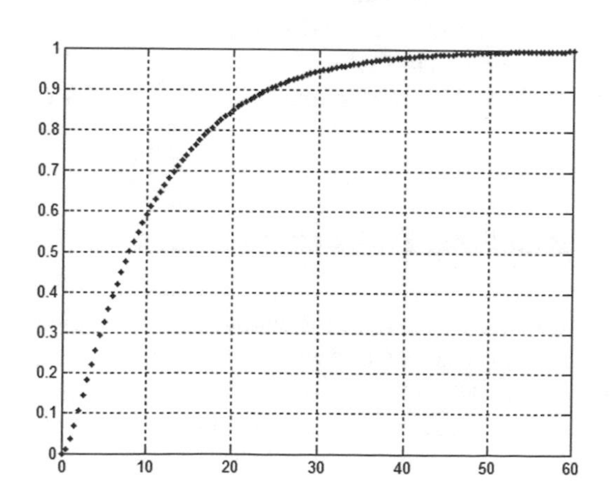

Fig. 4.4 Step response

(See formula (4.51) in the textbook [1]). Here A is the static gain and τ and T are the time constants of the numerator and the denominator, respectively. So $I_1 = 10 + 1 = 11$.

With MATLAB™ the value of the integral can be determined using the rectangle rule: $I_1 = [1 - y(0)]T_s + [1 - y(1)]T_s + \ldots + [1 - y(N)]T_s$.

Summation is executed by the command sum for the elements of the vector, which gives a good approximation to I_1.

```
I1 k = sum((1-y)*Ts)
     I1 k  = 11.2232
```

The result will be more accurate if the sampling time is smaller. Repeat the calculation for $T_s = 0.05$. The quadratic error integral can be evaluated similarly.

$$I_2 = \lim_{t\to\infty} \int_0^t e^2(\tau)\,\mathrm{d}\tau$$

```
I2 k = sum((1-y).*(1-y)*Ts)
     I2 k  =  6.2045
```

4.1.2 Demonstrating the Basic Properties of Negative Feedback

The effects of feedback can be described by the following properties:

(a) it modifies the transient behaviour;
(b) it improves reference signal tracking;
(c) it may stabilise an unstable process;
(d) it improves disturbance rejection;
(e) it improves the insensitivity to parameter changes;
(f) it has a linearizing effect;
(g) it can be used to approximate the inverse of a transfer function.

4.2 Resulting Transfer Functions

The usual block diagram of a control system is given in Fig. 4.5.

A filter $F(s)$ is not always applied. The disturbance sometimes is taken into consideration only at the output or at the input of the process. The behaviour of the system can be described by 6 different resulting transfer functions.

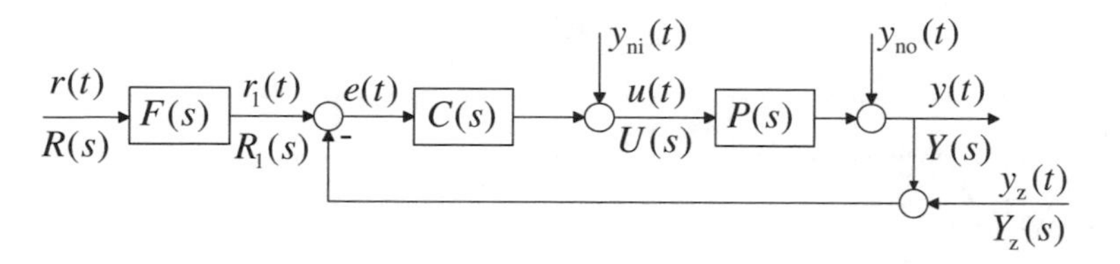

Fig. 4.5 Block diagram of a control system

Example 4.3 In a control system the process is given by the transfer function $P(s) = \frac{1}{(1+10s)(1+s)}$, the transfer function of the regulator is $C(s) = 5\frac{1+10s}{10s}$ and the filter is $F(s) = \frac{1}{1+s}$ (See the topic of regulator design in Sect. 8.2.3.).

Let us determine the following 6 resulting transfer functions:

$$\frac{Y(s)}{R(s)};\quad \frac{Y(s)}{Y_z(s)};\quad \frac{Y(s)}{Y_{ni}(s)}$$
$$\frac{U(s)}{R(s)};\quad \frac{U(s)}{Y_z(s)};\quad \frac{U(s)}{Y_{ni}(s)}$$

```
P = 1/((10*s + 1)*(1 + s))
C = 0.5*(0.5*s + 1)/s
F = 1/(1 + s)
L = minreal(C*P);
N = 1+L;
HYR = F*L/N;HYR = minreal(HYR)
HUR = F*C/N;HUR = minreal(HUR)
HYYz = -L/N;HYYz = minreal(HYYz)
HUYz = -C/N;HUYz = minreal(HUYz)
HYYni = P/N;HYYni = minreal(HYYni)
HUYni = 1/N;HUYni = minreal(HUYni)
H = [HYR,HYYz,HYYni;HUR,HUYz,HUYni];
```

Plot the step responses (Fig. 4.6) and the BODE diagrams. It can be seen that the different transfer functions cause different dynamics. The rejection of the effect of the step disturbance is faster than the dynamics of step reference signal tracking. In design, the dynamics of the reference signal tracking and the dynamics of the disturbance rejection can be influenced by the appropriate choice of the regulator $C(s)$ and the prefilter $F(s)$.

```
t = 0:0.1:10;
figure (1),step(H,t),grid on
figure (2),bode(H),grid on
```

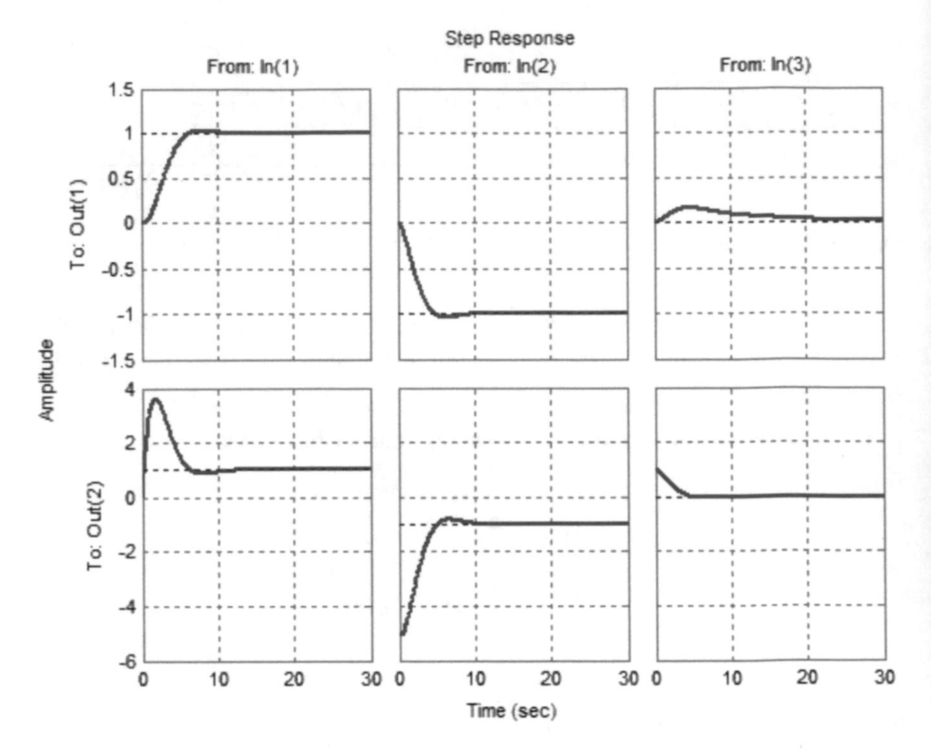

Fig. 4.6 Step responses of different signals in a control system

4.3 The Effect of the Poles of the Excitation Signal and the Effect of the Poles of the Open Loop on Steady State Behaviour

Let us consider an exponentially decreasing input signal with time constant $T_a = 10$, i.e. its pole is $p_a = -1/T_a = -0.1$. The input signal is $r(t) = e^{-t/T_a} = e^{-tp_a}$. Its LAPLACE transform is $R(s) = 1/(s - p_a)$. The open loop can be given by a second-order lag element with gain $K = 5$. Let us analyse the reference signal transfer properties of the closed loop, if the transfer function of the open loop contains the pole of the reference signal and has no smaller pole which would cause a slower response. Analyse the response also for the case when the open loop does not contain the pole of the reference signal.

```
s = zpk('s')
Ta = 10; K = 5
L = K/((Ta*s + 1)*(s + 1))
L1 = K/((2*Ta*s + 1)*(s + 1))
L2 = K/((0.5*Ta*s + 1)*(s + 1))
T = L/(1 + L);
T1 = L1/(1 + L1);
T2 = L2/(1 + L2);
t = 0:0.01:50;
r = exp(-t/Ta);
y = lsim(T,r,t);
y1 = lsim(T1,r,t);
y2 = lsim(T2,r,t);
figure (1); plot(t,r,'b',t,y,'r');
figure (2); plot(t,r,'b',t,y1,'r',t,y2,'g')
```

It can be seen that the output signal of the closed loop tracks accurately the input signal only if the transfer function of the open loop contains the pole of the reference signal (Figs. 4.7 and 4.8). A special case of tracking is when $p = 0$ is the pole of the reference signal, i.e. the input is the step reference signal, whose Laplace transform is $1/s$. In the transfer function of the open loop, a pole at zero represents an integrator. As in our example the open loop does not contain the pole of the step input (there is no integrator in the loop), there will be a steady state error (Fig. 4.9).

```
r = ones(1,length(t));
y = lsim(T,r,t);
figure (3); plot(t,r,'b',t,y,'r');
```

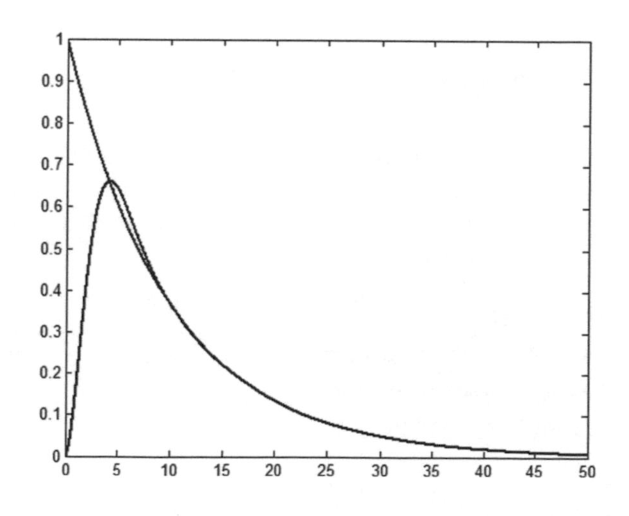

Fig. 4.7 Tracking of an exponential signal

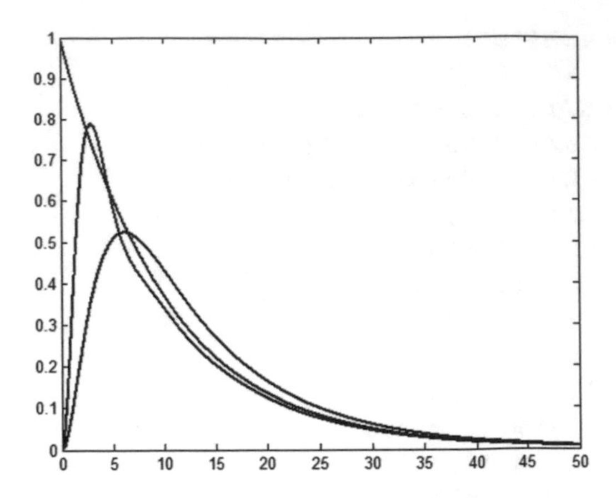

Fig. 4.8 Tracking of an exponential signal

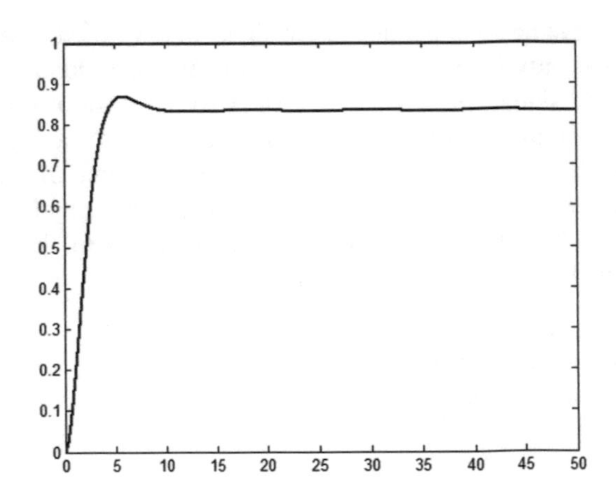

Fig. 4.9 Steady state error in step response

4.4 Properties of the Static Response

The steady state response of a feedback system (closed loop) depends on the type number of the system. 'Type number' means the number of integrators in the open loop. The table below gives the steady state error for different reference signals and different type numbers.

Type number	0	1	2
Unit step reference signal $j = 1$	$\frac{1}{1+K}$	0	0
Ramp reference signal $j = 2$	∞	$1/K$	0
Quadratic reference signal $j = 3$	∞	∞	$1/K$

Example 4.4 The loop transfer function of a system is

$$L(s) = \frac{K}{(1+s)(1+5s)} = \frac{10}{(1+s)(1+5s)}$$

Analyse the behaviour of the open- and the closed-loop.

```
s = zpk('s')
L = 10/((1 + s)*(5*s + 1))
T = feedback(L,1)
```

Determine the steady state values of the step responses of the open- and of the closed loop. According to the finite value theorem of the LAPLACE transformation for unit step input

$$r(t) = 1(t); \quad R(s) = \frac{1}{s}$$

$$y(t \to \infty) = \lim_{s \to 0} s\,R(s)\,H(s) = \lim_{s \to 0} s\frac{1}{s}H(s) = \lim_{s \to 0} H(s)$$

The steady state values of the step responses of the open- and of the closed-loop are then

$$y_{\text{open-loop}}(t \to \infty) = \lim_{s \to 0} L(s) = K = 10$$

$$y_{\text{closed-loop}}(t \to \infty) = \lim_{s \to 0} T(s) = \frac{y_{\text{open-loop}}(t \to \infty)}{1 + y_{\text{open-loop}}(t \to \infty)} = \frac{K}{1+K} = \frac{10}{1+10}$$

The value of the steady state error is: $e(t \to \infty) = 1 - y_{\text{open-loop}}(t \to \infty) = \frac{1}{1+K} = \frac{1}{11}$

Plot the step responses of the open- and the closed-loop in the same diagram:

```
step(L,'r',T,'b')
```

The steady values can be read from the diagrams or can be calculated with the command dcgain.

```
yos = dcgain(L)
ycs = dcgain(T)
```

Also plot the BODE diagrams of the open and the closed loop in the same diagram.

```
bode(L,'r',T,'b')
```

Determine the value of the steady state error for $K = 1, 20, 100$.

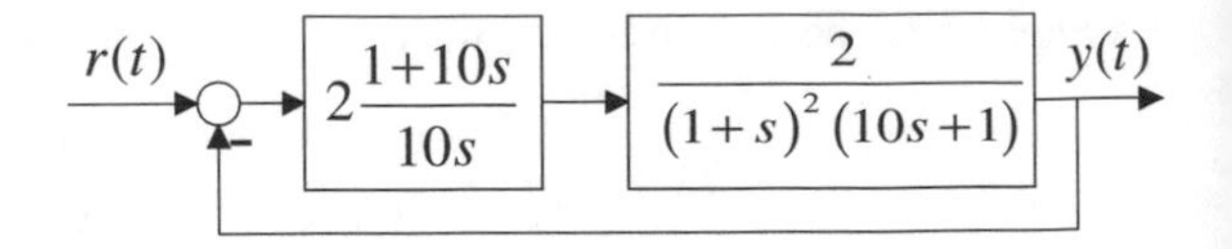

Fig. 4.10 Control system of type number 1

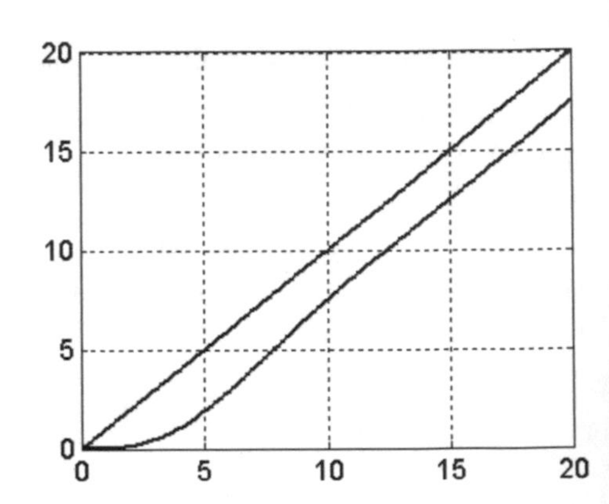

Fig. 4.11 The system tracks the ramp signal with steady error

Example 4.5 Determine the steady state error of the system given in Fig. 4.10 for unit step, ramp and parabolic reference signals.

The system contains one integrator, therefore its type number is 1. On the basis of the table above the steady error is

for step reference signal: $e(\infty) = 0$
for ramp reference signal: $e(\infty) = 1/\mathrm{K} = 1/(2 \cdot 2/10) = 2.5$
for parabolic reference signal: $e(\infty) = \infty$

Check the steady error value for a ramp reference signal by simulation (Fig. 4.11)!

```
s = zpk('s')
C = 2*(1 + 10*s)/(10*s)
P = 2/(s + 1)^2/(10*s + 1)
L = minreal(C*P)
T = L/(1 + L); T = minreal(T)
t = 0:0.1:20;
r = t;
y = lsim(T,r,t);
plot(t,r,t,y); grid
```

4.5 Relation Between the Frequency Functions of the Open- and Closed-Loop

The nonlinear relation $T(s) = L(s)/[1 + L(s)]$ describing the resulting transfer function of a control system based on negative feedback determines the behaviour of the control system. Let us analyse how this relation maps the complex plane L to the complex plane T. Plot the absolute value of the frequency function of the closed loop: $M = |L(j\omega)/[1+L(j\omega)]|$.

As -1 is a singularity of the mapping, the neighbourhood of this point is investigated. Write the following program as an m-file.

```
res = 0.01; Mlimit = 5;
x = -3:res:1;
y = -2:res:2;
Mm = zeros(length(y),length(x));
for kx = 1:length(x)
  for ky = 1:length(y)
    L = x(kx) + y(ky)*i;
    T = L/(1 + L);
    M = abs(T);
    if M > Mlimit M = Mlimit;
    end
    Mm(ky,kx) = M;
  end
end
surf(x,y, Mm), shading INTERP, colormap('jet'), view(0,90)
```

In window *figure* in menu '*tools*' with option '*rotate 3D*' the 3D surface can be visualised from an arbitrary viewpoint. The value of M should be restricted to ensure the visualisability of the picture. For fixed values of M, the contour lines are circles (Fig. 4.12).

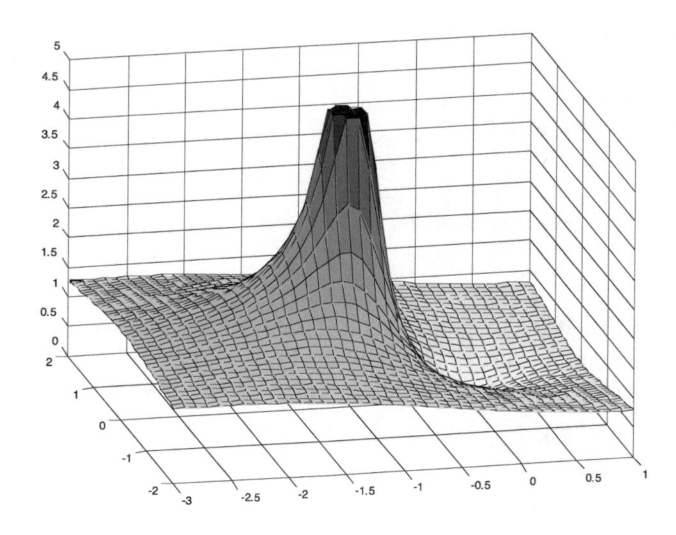

Fig. 4.12 Conform mapping of space L into space T

4.6 Relation Between the Overshoot of the Step Response and the Amplification of the Frequency Function

The maximum of the amplitude-frequency function of the closed-loop depends on how closely the Nyquist curve of the open-loop approaches the point -1 of the complex plane. On the Bode diagram this maximum means the amplification of the absolute value. Big amplification in the frequency domain means high overshoot in the time domain in the step response. The relationship between these values is not simple, as in the time domain the output signal is calculated by convolution, which means that the maximum overshoot in the time domain depends not only on the maximum value of the amplification in the frequency domain, but also on the amplifications on the other frequencies.

Example 4.6 Let us analyse, in the case of a second order oscillating element, how a change of the damping factor influences the overshoot in the time domain and the amplification of the amplitude in the frequency domain. Plot the values of the overshoot of the step response ym and the maximum amplitude of the frequency response Mm versus the damping factor (Fig. 4.13).

```
s = zpk('s')
T0 = 1;
kszi = [0.2:0.1:1];
t = 0:0.01:30;
w = logspace(-1,1,500);
for k = 1:length(kszi),
    T = 1/(s*s*T0*T0 + 2*T0*s*kszi(k) + 1);
    y = step(T,t);
    ym(k) = max(y);
    M = bode(T,w);
    Mm(k) = max(M);
    figure (2); hold on; plot(t,y)
    figure (3); loglog(w,M(:)); hold on;
end
figure (1);plot(kszi,ym,'r',kszi,Mm,'b'),
grid on
```

It can be seen that if the amplification in the frequency domain is higher, the overshoot in the time domain is also higher (Figs. 4.14 and 4.15). This holds exactly only for the given system and input signal, but shows well the important relations.

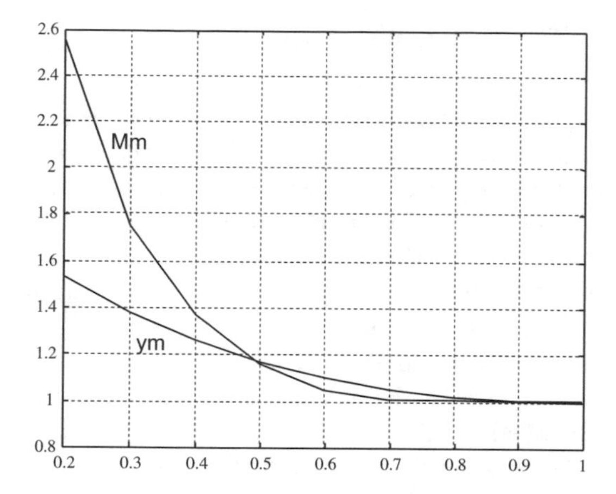

Fig. 4.13 The damping factor influences the overshoot and the BODE amplification

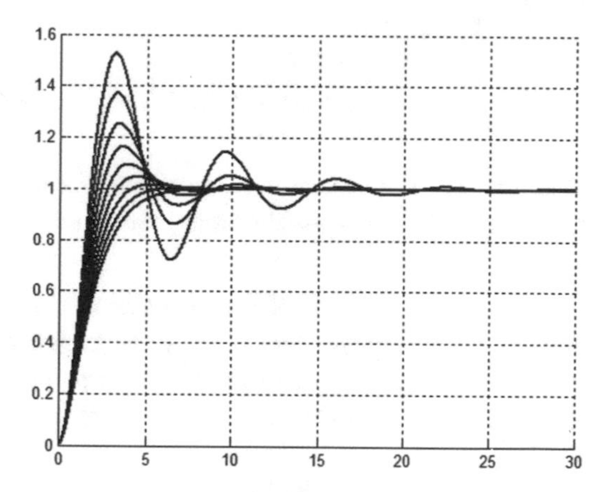

Fig. 4.14 Step responses of a second order oscillating element

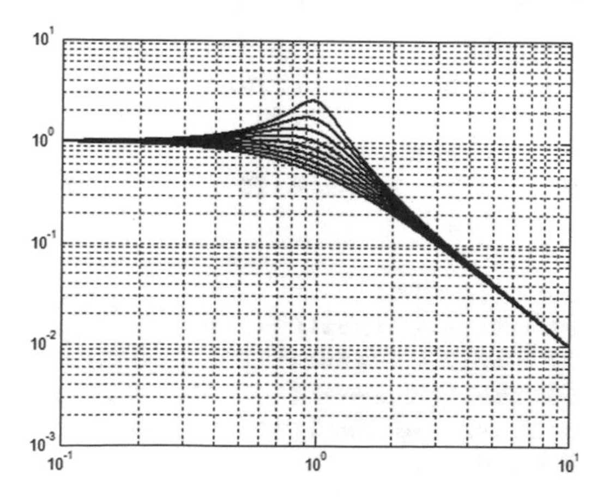

Fig. 4.15 BODE amplitude diagrams of a second order oscillating element

4.7 The Sensitivity Function

An important goal of a control system is to ensure acceptable behaviour also in case of changes in the parameters of the process model. In practical circumstances parameter changes can be the consequences of several effects. Warming of the system, ageing of its elements, change of humidity of the environment, etc., may influence significantly the behaviour of the system.

The effect of parameter changes can be investigated using the sensitivity function. The transfer function of the process can be given as the sum of the nominal transfer function and its change: $P(s) = P_o(s) + \Delta P(s)$. The sensitivity function S gives the ratio of the relative change of the resulting (overall) transfer function and the relative change of the transfer function of the process. So it characterises how much change is caused in the resulting transfer function if the process parameters change. The resulting transfer function T of the control system between the output and the reference signal is also called the complementary sensitivity function.

$$S = \frac{\Delta T/T}{\Delta P/P} = \frac{1}{1+CP}; \quad T = \frac{CP}{1+CP}; \quad S+T=1.$$

Example 4.7 The system is a second-order oscillating element with transfer function $P(s) = \frac{1}{1+2\xi T_1 s + s^2 T_1^2}$. Its time constant is $T_1 = 5$ and the damping factor is $\xi = 0.7$.

The transfer function of the regulator is $C(s) = \frac{1}{10s}$. Unity negative feedback is applied. Let us analyse how sensitive is the behaviour of the control system to the changes of the time constant and the damping factor. Let us analyse the dynamics of the control system if the time constant changes to $T_1 = 10$ and the damping factor to $\xi = 0.2$.

For both cases plot the Bode amplitude diagrams of the sensitivity function and of the relative change of the process in one diagram.

```
s = zpk('s')
w = logspace(-3,1,500);
T0 = 5;kszi0 = 0.7;
T1 = 10;kszi1 = 0.2;
P0 = 1/(1 + 2*kszi0*T0*s + T0^2*s^2)
C = 0.1/s
L0 = C*P0
S = 1/(1 + L0)
P1 = 1/(1 + 2*kszi0*T1*s + T1^2*s^2)
P2 = 1/(1 + 2*kszi1*T0*s + T0^2*s^2)
deltaP1 = minreal((P1-P0)/P0,0.001)
deltaP2 = minreal((P2-P0)/P0,0.001)
M = bode(S,w);
```

```
M1 = bode(deltaP1,w);
M2 = bode(deltaP2,w);
figure (1)
loglog(w,M(:),w,M1(:))
figure (2)
loglog(w,M(:),w,M2(:))
t = 0:0.1:100;
figure (3)
step(P0,t,P1,t)
figure (4)
step(P0,t,P2,t)
```

Figure 4.16 shows that decreasing the damping factor, the relative change of the process is significant in the frequency range where the sensitivity function shows also amplification. A closed-loop control system will react strongly to this change.

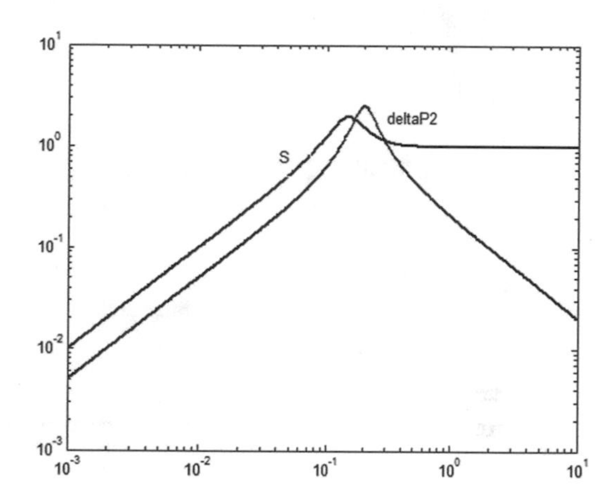

Fig. 4.16 Sensitivity function and the relative change in the damping factor

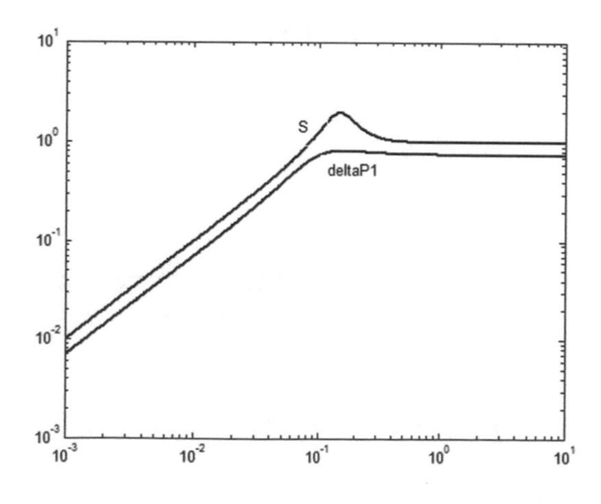

Fig. 4.17 Sensitivity function and the relative change in the time constant

The step response is shown in Fig. 4.18. Figure 4.17 shows the frequency function of the relative change of the process in the case of a change of the time constant. This curve is below the frequency function of the sensitivity function. The control system will be less sensitive to this parameter change (Fig. 4.19).

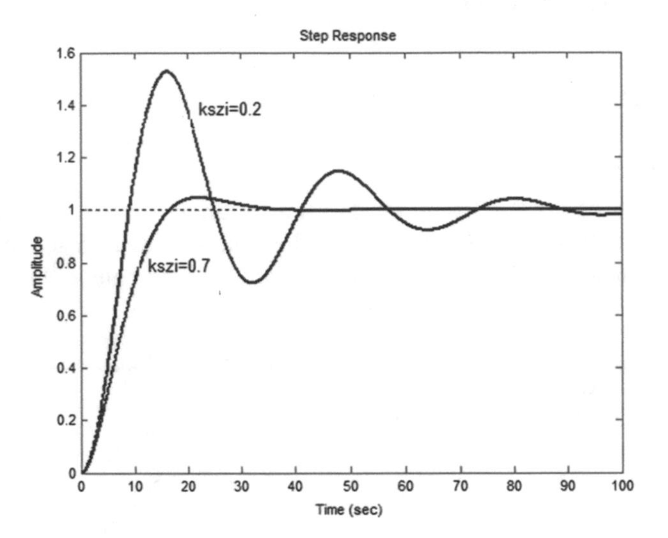

Fig. 4.18 Step responses in cases of two damping factors

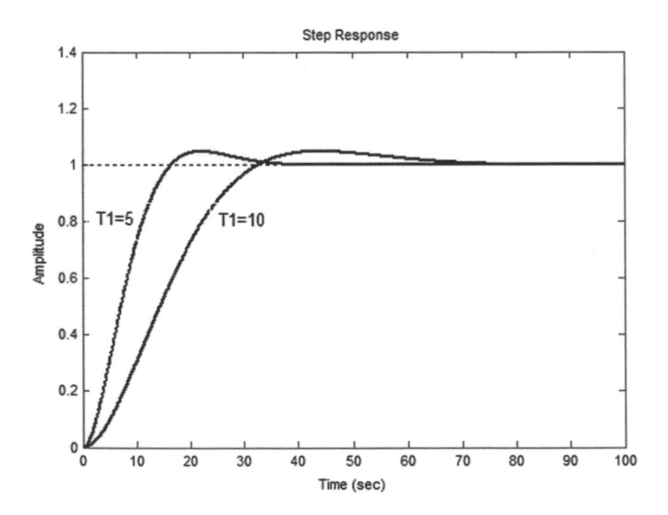

Fig. 4.19 Step responses in cases of two time constants

4.8 Control Structures

The most used control structure is realized by negative feedback where the regulator and the process are serially connected. This structure can be modified, supplemented with further elements to meet more sophisticated control aims (e.g. improvement of disturbance rejection).

4.8.1 *Feedforward*

Disturbance rejection can be improved if not only the output signal is used for control, but intermediate measurable signals are also employed to influence the control process. In these intermediate signals, the effect of the disturbance may show up earlier than in the output signal. In feedforward control, a measurable disturbance signal is measured and fed forward to influence the actuating signal.

Let us build a SIMULINK™ block-diagram to demonstrate feedforward control (Fig. 4.20).

Example 4.8 Let us simulate the behaviour of the system if $P = \frac{1}{(2s+1)(0.5s+1)}$, $C = \frac{2s+1}{s}$, $P_n = \frac{1}{s+1}$.

```
P = 1/((2*s + 1)*(0.5*s + 1))
C = (2*s + 1)/s
Pn = 1/(s + 1)
```

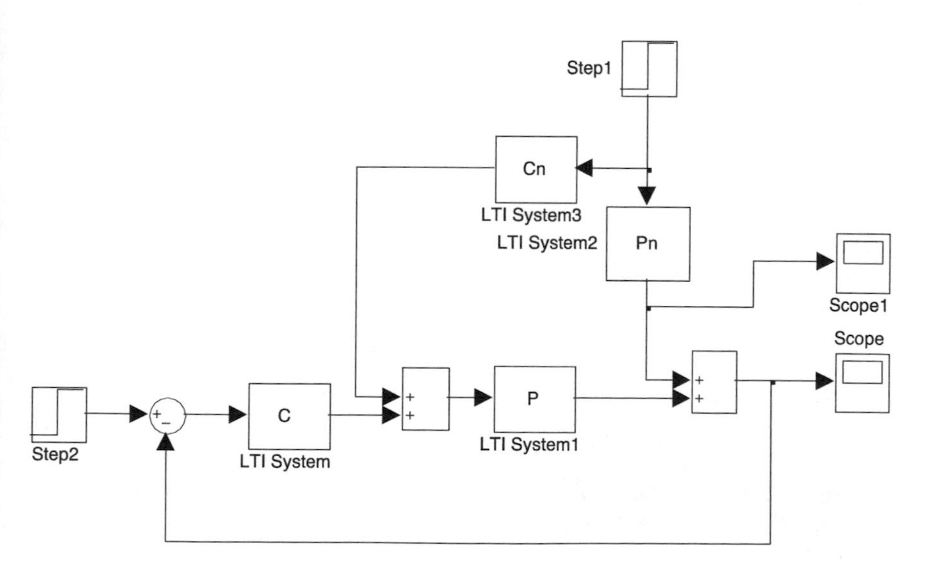

Fig. 4.20 SIMULINK™ block diagram of feedforward control

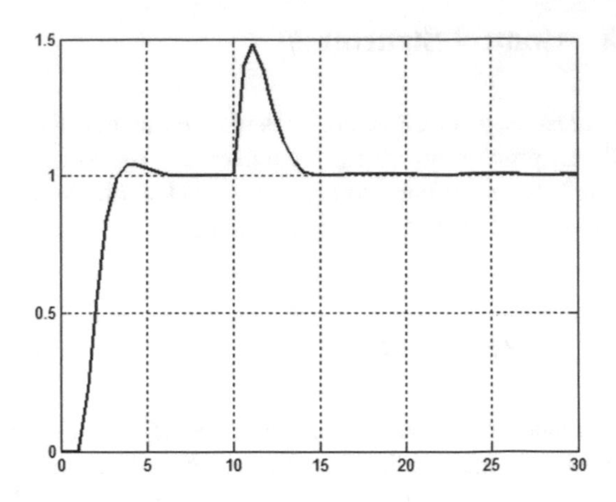

Fig. 4.21 Tracking and disturbance rejection without feedforward

The input signal is: $r(t) = 1(t)$.

The disturbance is: $y_n(t) = 1(t - 10)$.

Let us compare the behaviour of the control system without and with feedforward.

Without feedforward (Fig. 4.21):

```
Cn = 0
```

Feedforward is perfect if the effect of the disturbance through the feedforward regulator $C_n(s)$ cancels the effect of the disturbance, i.e. $P_n(s) + C_n(s)P(s) = 0$, hence $C_n(s) = -\frac{P_n(s)}{P(s)}$.

```
Cn = -Pn/P
- (s + 0.5) (s + 2)
--------------------
      (s + 1)
```

This transfer function is non-realizable as the degree of its numerator is higher than the degree of its denominator. Therefore an additional high frequency pole (a small time constant) is added to this transfer function.

$$C_n = -\frac{10(s+0.5)(s+2)}{(s+1)(s+10)}$$

```
Cn = -Pn/P/(0.1*s + 1)
```

Disturbance elimination will not be perfect, but the effect of the disturbance is decreased significantly (Fig. 4.22). Feedforward can be applied if the disturbance is measurable.

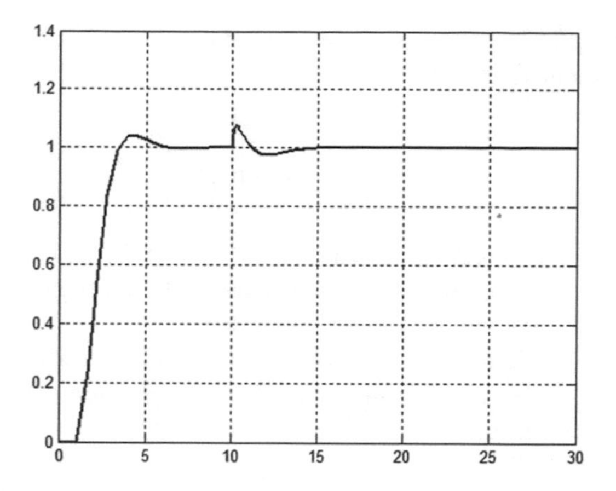

Fig. 4.22 Tracking and disturbance rejection with feedforward

4.8.2 Cascade Control

Cascade control can be applied if the process can be separated into several serially connected parts and the output signals of each part can be measured (Fig. 4.23).

For the first element of the system an inner control system can be built. For this inner circuit connected serially to the second part of the system an outer controller is designed (Fig. 4.24). The inner circuit can be fast, ensuring also fast disturbance rejection of the inner disturbance acting between the two parts of the system. The controller in the outer circuit can be designed for good reference signal tracking and rejection of the effect of the outer disturbance.

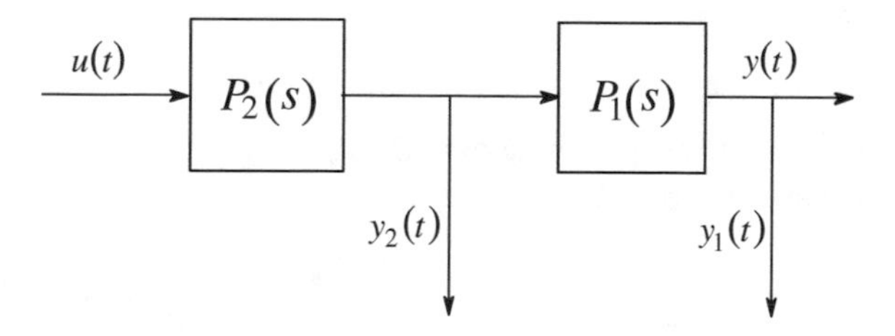

Fig. 4.23 A process separated to two serially connected parts with measurable inner signal

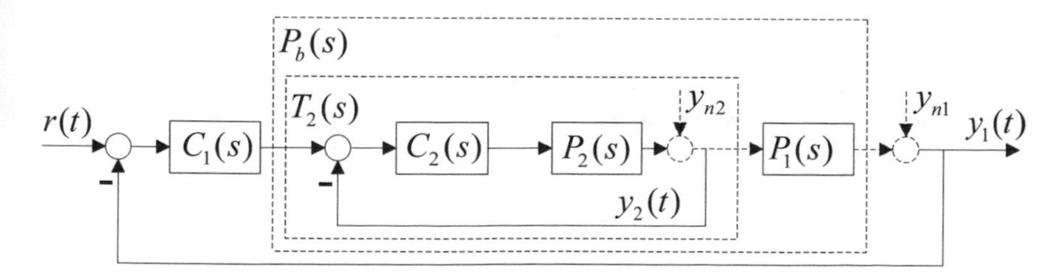

Fig. 4.24 Block diagram of cascade control

Example 4.9 The system to be controlled consists of two serially connected parts with transfer functions

$$P_1 = \frac{1}{1+10s} \quad \text{and} \quad P_2 = \frac{1}{1+s}.$$

The advantage of cascade control is significant if the system consists of a faster and a slower part and the slower part with the bigger time constant is in the outer circuit. Here this condition is fulfilled.

Let us design a fast control in the inner circuit, which ensures fast rejection of the effect of the inner disturbance. Then design a regulator for the outer circuit which ensures tracking of the step reference signal without steady error and rejection of the outer disturbance.

In the inner control circuit, let us choose a proportional regulator (C2 = 10).

```
P2 = 1/(s + 1)
C2 = 10
T2 = C2*P2/(1 + C2*P2);T2 = minreal(T2)
step(P2,'b',T2,'r'),grid
```

On the left side of Fig. 4.25 it can be seen that the behaviour of the inner circuit has become fast, but there is a static error.

In the outer control circuit an integrator has to be used in the regulator to decrease the steady state error to zero. With a regulator C1 the big time constant of the *PI* part of the system is cancelled and instead an integrating effect is introduced. Let $C_1(s) = \frac{5(1+10s)}{s}$. (Regulator design using considerations in the frequency domain, the so called *PID* compensation is discussed in Chap. 8.)

```
P1 = 1/(10*s + 1)
Pb = T2*P1
C1 = 5*(10*s + 1)/s
T = C1*Pb/(1 + C1*Pb); T = minreal(T)
figure (1),step(T),grid
```

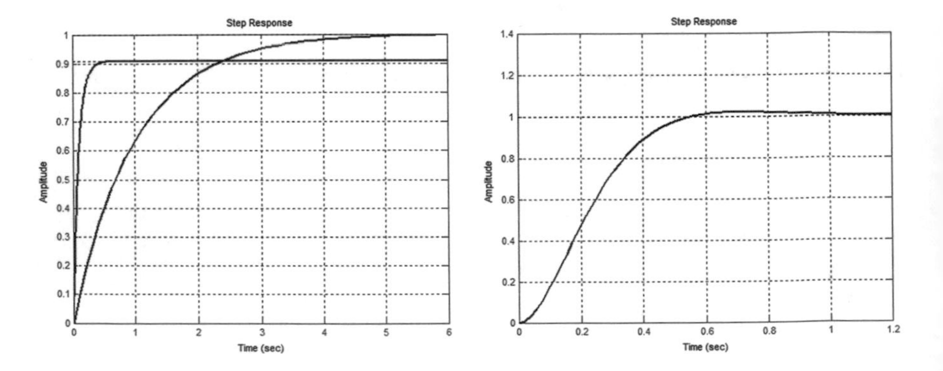

Fig. 4.25 Behaviour of the inner loop and of the output signal in cascade control

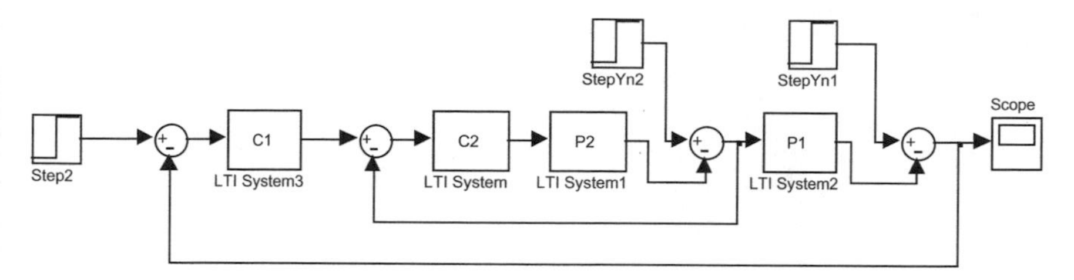

Fig. 4.26 SIMULINK™ block diagram of cascade control

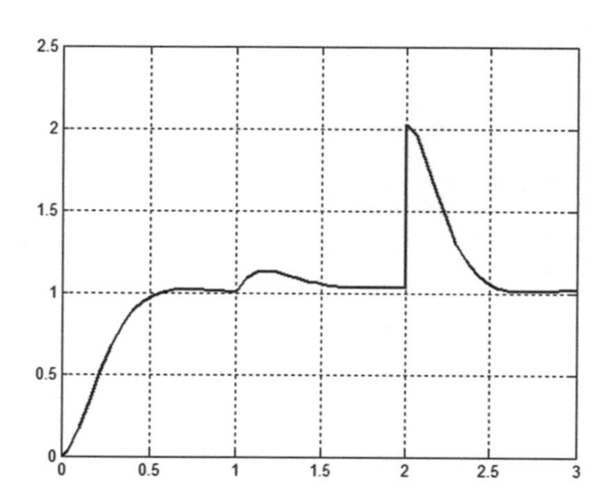

Fig. 4.27 Simulation of cascade control for tracking and disturbance rejection

On the right side of Fig. 4.25 it can be seen that the output response is fast and the static error is zero.

Let us investigate the behaviour of the cascade control circuit in the SIMULINK™ environment (Fig. 4.26).

Set the reference signal and the disturbances to the following values:

$$r(t) = 1(t);\; y_{n2} = 1(t-1);\; y_{n1} = 1(t-2)$$

Set the simulation time to 3 s. In Fig. 4.27 it can be seen that the control system tracks the reference signal and eliminates the effects of both the inner and the outer disturbances.

Chapter 5
Stability of Linear Control Systems

A general notion of stability says that a system is stable if after being removed from a stable state, the system returns to the original state provided no external input is applied. Another general notion of stability is called *BIBO* stability, i.e. bounded output is obtained as a response to any and all bounded inputs.

Consider the closed-loop system given in Fig. 5.1. The open-loop transfer function is $L(s)$ and a unity feedback is applied around $L(s)$.

The resulting transfer function of the closed-loop system is:

$$T(s) = \frac{L(s)}{1 + L(s)}$$

It is well known that all the components of the transient response will decay once the roots of the closed-loop characteristic equation $1 + L(s) = 0$ are located in the left half-plane side of the complex plane. The characteristic equation contains the denominator polynomial of the closed-loop transfer function above. Consequently, the roots of the characteristic equation are identical to the poles of the closed-loop system. Note that using state-space representations, this statement is only valid for systems which are both controllable and observable.

5.1 *BIBO* Stability

The *BIBO* stability criterion can be used to check the stability of linear systems. In practice a natural choice is to apply a unit step excitation as a bounded input.

Example 5.1.1 Assume we have the following closed-loop transfer function:

$$T(s) = \frac{s + 5}{s^5 - 3s^4 + 4s^3 + 10s^2 + 5s - 10}.$$

L. Keviczky et al., *Control Engineering: MATLAB Exercises*,
Advanced Textbooks in Control and Signal Processing,
https://doi.org/10.1007/978-981-10-8321-1_5

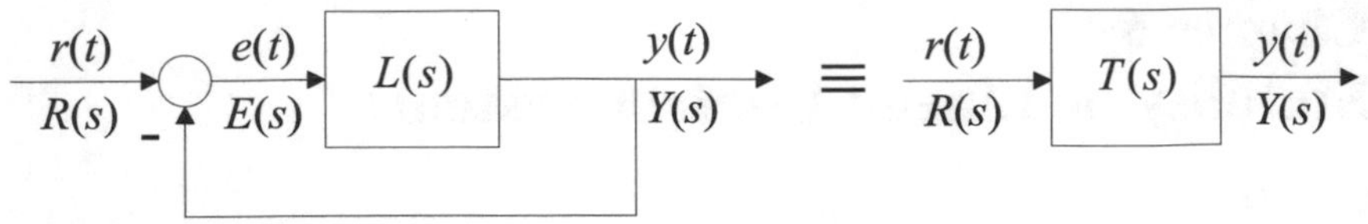

Fig. 5.1 Block diagram of a closed-loop control system

Applying a unit step input, check the stability of the closed-loop system. Use MATLAB™ commands, such as

```
num=[1, 5]
den=[1, -3, 4, 10, 5, -10]
T=tf(num,den)
step(T)
```

It can be seen that the step response will not be bounded, meaning that the closed-loop system is unstable.

5.2 Stability Analysis Based on the Location of the Closed-Loop Poles

One way to obtain the system output in analytical form is to derive the partial fraction expansion form of the Laplace transform of the transfer function. Then the analytical solution of the output signal in the time domain is simply obtained by performing an inverse Laplace transformation. In more detail, the Laplace transform of the output signal is the sum of the components that take the form $\frac{r_i}{s-p_i}$, where p_i denotes a system pole. Consequently, the system's stability can be determined based on the location of the system poles. A system turns out to be stable once all the poles are located in the left half-plane. Moreover, it can be seen whether oscillating components are expected to show up, as is indicated by the existence of complex conjugate poles in the left half-plane.

Example 5.2.1 Check the stability of the system introduced in Example 5.1.1. Check the location of the poles in this analysis:

```
poles=roots(den)
          poles =
           2.1150 + 2.1652i
           2.1150 - 2.1652i
          -0.9824 + 0.7214i
          -0.9824 - 0.7214i
           0.7348
```

or in another form

```
[zeros,poles,KonstGain]=zpkdata(T,'v')
```

Either way, it can be seen that the system has poles in the right half-plane.

Alternatively, the pzmap command shows the location of both the zeros and poles in graphical mode:

```
pzmap(T)
```

Note that complex poles produce oscillations in the output response. However, this can only barely be seen because of the dominant exponential growth.

5.3 Stability Analysis Using the Routh-Hurwitz Criterion

The poles of the closed-loop characteristic equation can be calculated analytically only for polynomials of degree less than five. If MATLAB™ is not available, higher degree equations need to be solved, which can only be solved numerically. If dead-time is included in the open-loop, the equation takes a transcendental form, causing further difficulties for the solution.

If the system is free of dead-time, methods have been developed to judge the stability based on relations between the roots and coefficients of the characteristic equation.

In the sequel assume that the closed-loop characteristic equation is given in the following form:

$$\mathcal{A}(s) = a_n s^n + a_{n-1} s^{n-1} + \ldots + a_1 s + a_o = a_n (s - p_1)(s - p_2)\ldots(s - p_n) = 0$$

5.3.1 *Stability Analysis Using the Routh Scheme*

Set up the following (so called Routh scheme) from the coefficients of the characteristic equation:

$$\begin{matrix} a_n & a_{n-2} & a_{n-4} & a_{n-6} & \cdots \\ a_{n-1} & a_{n-3} & a_{n-5} & a_{n-7} & \cdots \\ b_{n-2} & b_{n-4} & b_{n-6} & b_{n-8} & \cdots \\ c_{n-3} & c_{n-5} & c_{n-7} & c_{n-9} & \cdots \\ \vdots & & & & \end{matrix}$$

where

$$b_{n-2} = \frac{a_{n-1}a_{n-2} - a_n a_{n-3}}{a_{n-1}}, \quad b_{n-4} = \frac{a_{n-1}a_{n-4} - a_n a_{n-5}}{a_{n-1}}, \quad b_{n-6} = \frac{a_{n-1}a_{n-6} - a_n a_{n-7}}{a_{n-1}}, \ldots$$

$$c_{n-3} = \frac{b_{n-2}a_{n-3} - a_{n-1}b_{n-4}}{b_{n-2}}, \quad c_{n-5} = \frac{b_{n-2}a_{n-5} - a_{n-1}b_{n-6}}{b_{n-2}}, \ldots$$

It can be seen that the length of the consecutive rows is getting shorter and shorter. For a characteristic equation of order n, the scheme consists of $n + 1$ rows. Elements with negative indices should be interpreted as elements whose value is zero.

The system is stable if all the coefficients of the characteristic equation are positive and all the elements in the first (leftmost) column of the Routh scheme are also positive. If there are changes in sign along the first column, the number of the sign changes equals the number of poles in the right half-plane (i.e. the number of unstable poles).

Example 5.3.1 The transfer function of a loop transfer function is

$$L(s) = \frac{K}{(1+10s)(1+5s)(1+s)(1+0.5s)}.$$

Find the critical value of the gain K (loop gain) yielding a stable closed-loop system. Consider first the characteristic equation. Start with defining $L(s)$:

```
s=tf('s')
den=(1+10*s)*(1+5*s)*(1+s)*(1+0.5*s)
     25 s^4 + 82.5 s^3 + 73 s^2 + 16.5 s + 1
```

The closed-loop characteristic equation becomes

$$1 + L(s) = 0 = 25s^4 + 82.5s^3 + 73s^2 + 16.5s + 1 + K$$

The coefficients in the Routh scheme are:

$$a_4 = 25; \quad a_3 = 82.5; \quad a_2 = 73; \quad a_1 = 16.5; \quad a_o = 1 + K;$$

$$b_2 = \frac{a_3 a_2 - a_4 a_1}{a_3} = \frac{82.5 \cdot 73 - 25 \cdot 16.5}{82.5} = 68;$$

$$b_o = \frac{a_3 a_o - a_4 a_{-1}}{a_3} = \frac{82.5(1+K) - 0}{82.5} = 1 + K;$$

$$c_1 = \frac{b_2 a_1 - a_3 b_o}{b_2} = \frac{68 \cdot 16.5 - 82.5(1+K)}{68} = 16.5 - 1.2132(1+K); \text{ and}$$

$$d_o = \frac{c_1 b_o - b_2 c_{-1}}{c_1} = b_o = 1 + K.$$

The ROUTH scheme can then be constructed:

$$\begin{array}{ccc} 25 & 73 & 1+K \\ 82.5 & 16.5 & \\ 68 & 1+K & \\ 16.5-1.2132(1+K) & 0 & \\ 1+K & & \end{array}$$

For stability, all the values in the first column must be positive. This is fulfilled when $-1<K<12.6$.

5.3.2 Stability Analysis Based on the HURWITZ Determinant

Using the coefficients of the characteristic equation construct the following (so called HURWITZ) determinant:

$$\begin{vmatrix} a_{n-1} & a_{n-3} & a_{n-5} & a_{n-7} & \cdots \\ a_n & a_{n-2} & a_{n-4} & a_{n-6} & \cdots \\ 0 & a_{n-1} & a_{n-3} & a_{n-5} & \cdots \\ 0 & a_n & a_{n-2} & a_{n-4} & \cdots \\ 0 & 0 & a_{n-1} & a_{n-3} & \cdots \\ \vdots & & & & \cdots \end{vmatrix}$$

Elements with negative indices will be taken to have the value zero.

The system is stable if all the coefficients of the characteristic equation are positive and all the sub-determinants along the main diagonal of the HURWITZ determinant are positive: $\Delta_i > 0$.

Example 5.3.2 Solve the problem discussed in Example 5.3.1 using the method of the HURWITZ determinant. Start with building up the HURWITZ determinant:

$$\begin{vmatrix} 82.5 & 16.5 & 0 & 0 \\ 25 & 73 & 1+K & 0 \\ 0 & 82.5 & 16.5 & 0 \\ 0 & 25 & 73 & 1+K \end{vmatrix}$$

The sub-determinants along the main diagonal are

$$\Delta_1 = 82.5 > 0; \quad \Delta_2 = 82.5 \cdot 73 - 16.5 \cdot 25 = 5610 > 0;$$
$$\Delta_3 = -(1+K)82.5^2 + 16.5 \cdot 5610 > 0; \text{ and}$$
$$\Delta_4 = (1+K)\Delta_3 > 0$$

The stability conditions from Δ_3 and Δ_4 are directly read: $-1 < K < 12.6$

Note that $K = -1$ would mean positive feedback and this result is obtained also for $a_o > 0$.

An additional problem:

In a closed-loop control system, the open-loop transfer function is

$$L(s) = \frac{K}{(1+sT_1)(1+sT_2)(1+sT_3)}.$$

(a) Find the critical value (maximum for closed-loop stability) of the gain K, if $T_1 = 1$, $T_2 = 0.4$, $T_3 = 0.1$.
(b) Find the critical value of K, if $T_1 = T_2 = T_3 = T$.
(c) What pair of K and T_3 guarantee closed-loop stability if $T_1 = 1$ and $T_2 = 0.4$?

Plot the function $K_{\text{krit}} = f(T_3)$. Show that if $T_3 \to 0$ or if $T_3 \to \infty$, the closed-loop system remains stable even for an infinitely large loop gain.

Solve this problem using either the ROUTH scheme or the HURWITZ determinant.

5.4 Stability Analysis Based on the Root-Locus Method

The root-locus method is a grapho-analytical method to show the poles of the closed-loop system as one parameter (typically the loop gain) in the system varies from zero to infinity. Note that a zero loop gain means an open-loop system.

If the poles of the characteristic equation are sitting on the imaginary axis, the closed-loop system is just about to be unstable (borderline stability). The root-locus method can not only determine closed-loop stability, but can also yield information on the dynamics of the closed-loop system. Root-locus points on the negative real axis suggest aperiodic transients in the time domain, and root-locus stages with complex poles in the left half-plane indicate oscillatory behaviour with damping.

MATLAB™ offers the rlocus command to draw the root-locus. Another MATLAB™ command (rlocfind) is to be used to find the gain belonging to a given point of the root-locus. rlocfind puts up a crosshair cursor in the graphics window which is used to select a pole location on an existing root locus.

Example 5.4.1 Draw the root-locus of a system given by the open-loop transfer function

$$L(s) = \frac{K}{(1+10s)(1+5s)(1+s)(1+0.5s)}$$

and find the critical value of the loop gain.

Set up the system with $K = 1$, then draw the root-locus. Then read the gain at the point where the root-locus is crossing the imaginary axis.

```
s=zpk('s')
L=1/((1+10*s)*(1+5*s)*(1+s)*(1+0.5*s))
rlocus(L)
rlocfind(L)
```

Now a left click on the critical point provides the value of the critical gain. Also, the exact coordinates of the selected point are shown. To derive an appropriate result it is worthwhile to zoom on the vicinity of the critical point before rlocfind is employed.

```
selected_point =
    0.0003 + 0.4466i
    ans = 12.5753
```

The root-locus is shown in Fig. 5.2. Show the step response of the closed-loop system at the critical value of the loop gain:

```
K=ans
t=0:0.01:40;
step(K*L/(1+K*L),t)
```

It can be seen that the system output exhibits oscillations with constant amplitude.

Example 5.4.2 Consider a system with the following loop transfer function:

$$L_1(s) = \frac{k}{s(s+2)(s+4)}$$

Sketch the root-locus and find the critical loop gain. Then study the root-locus after an additional zero is introduced in the loop transfer function.

$$L_{2,3}(s) = \frac{k(s+\alpha)}{s(s+2)(s+4)},$$

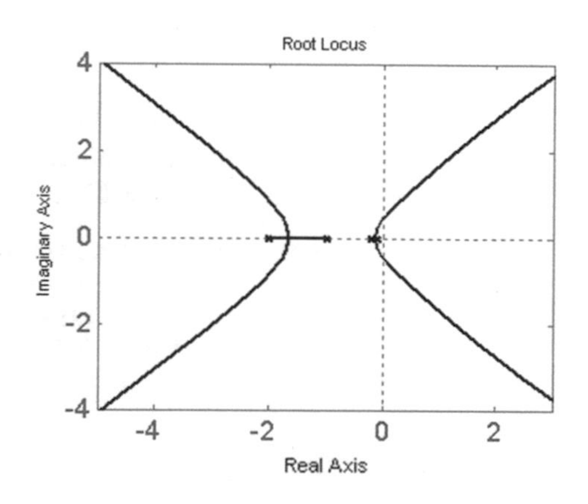

Fig. 5.2 Root-locus of a fourth-order system

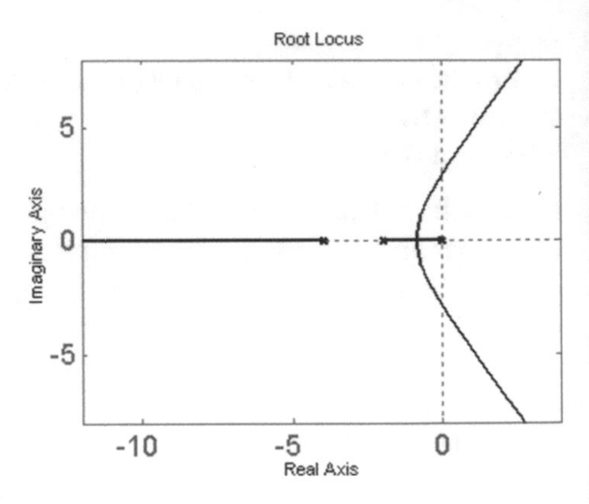

Fig. 5.3 Root-locus of a third-order system

where α takes the values of 3 or 1.

Discuss the stability issues of the extended system.

```
L1=1/(s*(s+2)*(s+4))
L2=L1*(s+3)
L3=L1*(s+1)
figure(1),rlocus(L1)
figure(2),rlocus(L2)
figure(3),rlocus(L3)
```

The root-locus for L_1 is shown in Fig. 5.3. To find the critical value of the loop gain zoom and use the command

```
rlocfind(L1)
```

Apply '*Select a point in the graphics window*' offered by MATLAB™, which results in

```
selected_point =
   0.0006 + 2.8252i
ans =
   47.9006
```

This critical value of k just obtained can also be checked by analytical tools.

It is seen that the inserted zero attracts one branch of the root-locus and the closed-loop becomes structurally stable in both cases ($L_{2,3}$ in Figs. 5.4 and 5.5).

Example 5.4.3 Consider the open-loop transfer function:

$$L(s) = \frac{k(s+4)(s+6)}{s(s+2)(s+8)}$$

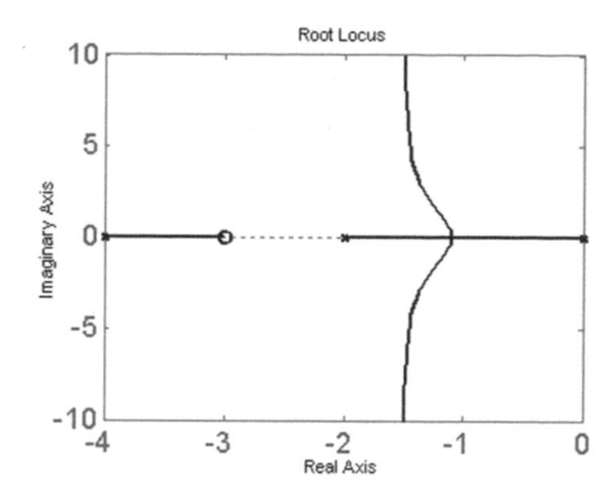

Fig. 5.4 Root-locus of a third-order system with a zero

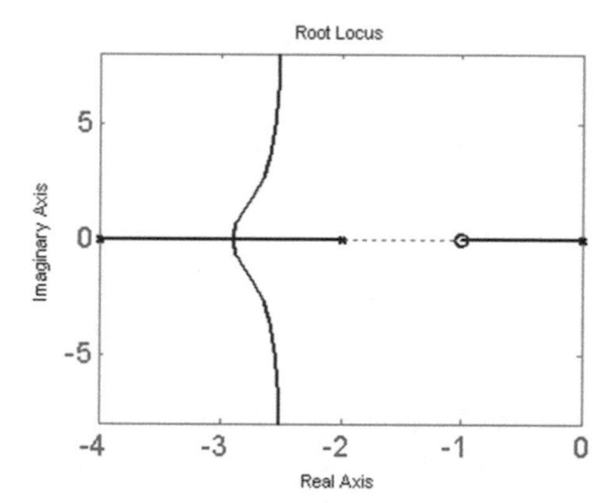

Fig. 5.5 Root-locus of a third-order system with a zero

Sketch the root-locus and evaluate the dynamic behaviour of the closed-loop system:

```
L=((s+4)*(s+6))/(s*(s+2)*(s+8))
rlocus(L)
```

The root-locus is shown in Fig. 5.6. Crossing the real axis can be obtained using rlocfind(L): -1.2 and -4.8865, and the corresponding loop-gain values are 0.4857 and 44.48, respectively. The closed-loop system is structurally stable, specifically for $0.4857 < k < 44.48$ the transient response will be determined by the complex conjugate closed-loop poles. Otherwise the transient response is aperiodic. Having two zeros and three poles, as the loop-gain grows to infinity one branch of the root-locus tends to go to infinity and the other two will converge to the finite zeros. Also, a point on the real axis is part of the root-locus once the total number of the zeros and poles located to the right from this point of the real axis is odd. A set of these point defines a complete region of the real axis.

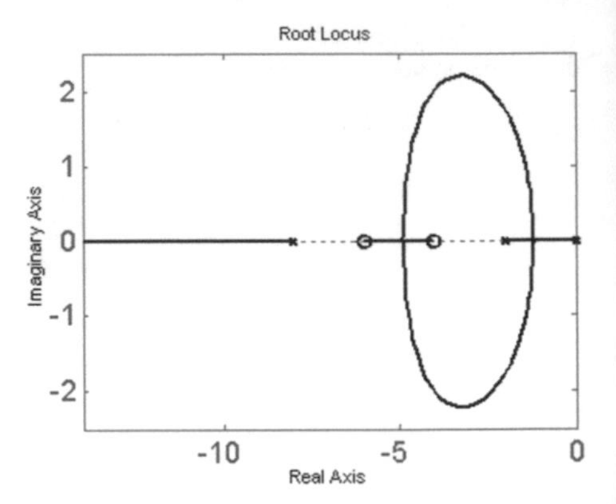

Fig. 5.6 Root-locus of a system with 3 poles and 2 zeros

Example 5.4.4 Let the transfer function of the loop transfer function be: $L(s) = \frac{k(s+2)}{(s-3)(s+5)(s+8)}$.

The open-loop is unstable as a consequence of an open-loop pole in the right half-plane. Can we stabilize the closed-loop system by applying feedback with a proper loop gain?

Draw the root-locus first:

```
L=(s+2)/((s-3)*(s+5)*(s+8))
rlocus(L)
```

The root-locus in Fig. 5.7 shows that all the poles of the closed-loop system will be in the left half-plane if the loop gain exceeds a certain (critical) value. Clearly, the closed-loop system can be stabilized once a sufficiently large loop gain is selected. The critical value of the loop gain can be determined using rlocfind(L) just as before. The critical loop gain will turn out to be 60.

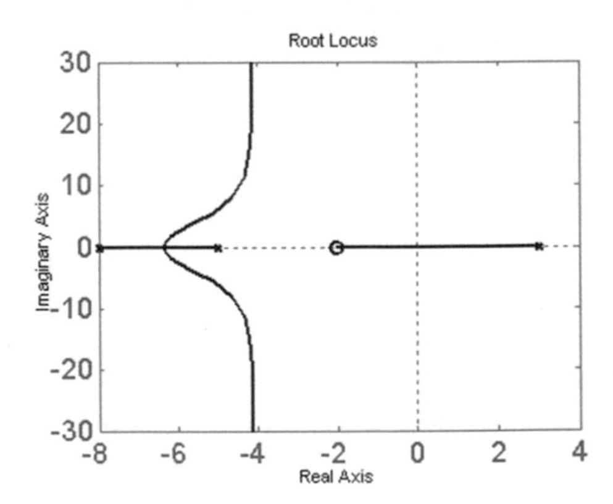

Fig. 5.7 Root-locus of a system with an unstable pole

Note that several root-locus curves can be drawn in the same figure. The root-locus can be drawn also for gain values given in a vector. The root-locus points and the coherent gains also can be obtained. The MATLAB™ commands are

```
rlocus(L,K)
rlocus(L1,L2,...)
rlocus(L1,'r',L2,'g:',L3,'mx')
[R,K]=rlocus(L)
```

5.5 Nyquist Stability Criterion

Having designed a closed-loop control system, stability is the most important attribute to be checked. Typically, we check the stability of a closed-loop system based on the behaviour of the open-loop system. Several methods exist to perform this step. The Nyquist stability criterion clarifies the stability issues of the closed-loop system given by $T(s) = \frac{Y(s)}{R(s)} = \frac{L(s)}{1+L(s)}$ based on the analysis of the frequency function of the open-loop transfer function $L(s) = \frac{Y(s)}{E(s)}$ (Fig. 5.8).

(a) To check the closed-loop stability, a simplified version of the Nyquist criterion can be employed if the open-loop transfer function has no unstable pole (pole with positive real part). The closed-loop system is then stable if the complete Nyquist diagram of the open-loop system does not encircle the point $(-1+0j)$ of the complex plane.

(b) To check the closed-loop stability, the generalized Nyquist criterion is to be employed if the open-loop transfer function has unstable poles (poles with positive real part). The closed-loop system is then stable if the number times $(-1+0j)$ is encircled by the complete Nyquist diagram of the open-loop system is equal to the number of unstable poles of the open-loop transfer function. The number of times a point encircled (*the winding number*) is considered to be positive when the path is traversed counter-clockwise. Note that the simplified Nyquist criterion is a special case of the generalized Nyquist criterion.

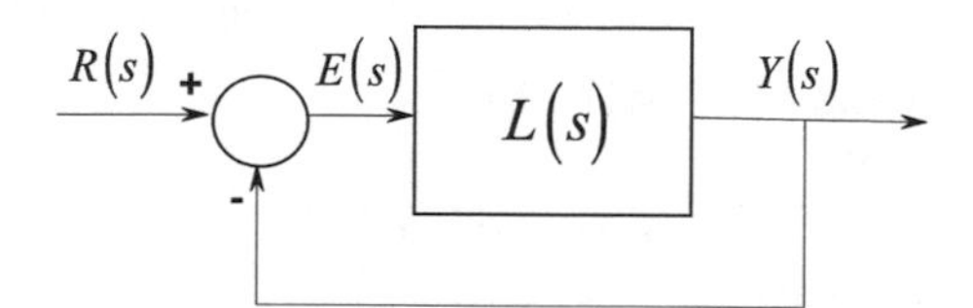

Fig. 5.8 Scheme of a closed-loop system

5.5.1 The Simplified Nyquist Stability Criterion

Example 5.5.1 Assume the open-loop transfer function is given by $L(s) = \frac{10}{(1 + 10s)(1+s)}$.
Use negative unity feedback. Check the stability of the closed-loop system using the simplified Nyquist criterion:

```
s=zpk('s')
L=10/((1+10*s)*(1+s))
[z,p,k]=zpkdata(L,'v')
```

It can be seen that the open-loop system has no unstable pole, thus the simplified Nyquist criterion is applicable.

```
nyquist(L),grid
```

The Nyquist diagram does not encircle the point $(-1+0j)$, so the closed-loop system is stable. Furthermore, we can conclude that the closed-loop system is structurally stable.

5.5.2 The Generalized Nyquist Stability Criterion

Example 5.5.2 Suppose given the open-loop transfer function $L(s) = \frac{-5}{(1-10s)(1 + 0.1s)}$.
Use negative unity feedback. Check the stability of the closed-loop system using the generalized Nyquist criterion:

```
s=zpk('s')
L=-5/((1-10*s)*(1+0.1*s))
[z,p,k]=zpkdata(L,'v')
```

Note that the open-loop system is unstable $(p_2 = 0.1)$.

```
nyquist(L)
```

The complete Nyquist diagram winds around $(-1+0j)$ in the positive sense (i.e., counter-clockwise). Consequently, the closed-loop system is stable.

Check this result by calculating the closed-loop poles:

```
T=feedback(L,1)
step(T)
[z,p,k]=zpkdata(T,'v')
pzmap(T)
```

Repeat this analysis when changing the sign of the open-loop poles:

$$L(s) = \frac{-5}{(1 + 10s)(1 - 0.1s)}$$

Look how the NYQUIST diagram will encircle $(-1 + 0j)$. Will the closed-loop system be stable?

Example 5.5.3 Consider the open-loop transfer function $L(s) = k\frac{1-s}{(1 + s)(1 + 0.5s)}$.

Negative unity feedback is applied. Find those values of the loop gain k which results in stable closed-loop system.

(a) To start with, assume $k = 1$:

```
L=(1-s)/((1+s)*(1+0.5*s))
[z,p,k]=zpkdata(L,'v')
```

All poles being stable allows us to use the simplified NYQUIST criterion.

```
nyquist(L); grid
```

Find k as the NYQUIST diagram crosses the real axis $(k = -0.666)$. Apply the zoom command or use the zoom option from the menu to read off this value. Increasing k will magnify the NYQUIST plot in the sense that all the points of the NYQUIST diagram will have an increased distance from the origin. The closed-loop system comes to borderline stability if the NYQUIST diagram crosses the real axis at -1. To achieve this, $k = 1/0.666 = 1.5$ is to be applied. So the closed-loop system will be stable for $0 < k < 1.5$ (if k is positive).

(b) Assume $k = -1$:

```
nyquist(-L), grid
```

Find again the stability region as before. Here $k > -1$ will be obtained.

Summing up the two conditions, we have $-1 < k < 1.5$ for the closed-loop stability.

5.6 Phase Margin, Gain Margin, Modulus Margin, Delay Margin

Beyond the fact that a system is stable, we are also interested in seeing how far we are from the borderline of stability. Several measures exist to characterize how far a stable system is from being unstable.

5.6.1 Phase Margin, Gain Margin

The phase margin defines the value of the phase angle needed to decrease the phase at the cut-off frequency to achieve borderline stability. The phase margin can be expressed analytically:

$$\varphi_t = \varphi(\omega_c) + 180^\circ$$

where ω_c is the cut-off frequency defined by

$$\omega_c \text{ is such that } |L(j\omega)|_{\omega=\omega_c} = 1$$

A closed-loop system is stable if the phase margin is positive. For example, if $\varphi = -120^\circ$ at the cut-off frequency, the phase margin is $\varphi_t = \varphi(\omega_c) + 180^\circ = -120^\circ + 180^\circ = 60^\circ$. This means that the closed-loop system is stable. For design purposes, 60° for the phase margin is a typical prescription.

The gain margin g_t is the factor by which the loop gain is to be multiplied to push a closed-loop system to borderline stability, i.e.

$$g_t = \frac{1}{|L(\omega_\pi)|}, \quad \text{where } \omega_\pi\text{: } \varphi(\omega)_{\omega=\omega_\pi} = -180^\circ$$

Here ω_π is that frequency where the phase angle is -180°. The closed-loop system is stable if the gain margin exceeds 1. A reasonable design prescription for the gain margin is around 2. MATLAB™ offers the margin command both to calculate and plot the phase and gain margin values.

Example 5.6.1 Given the open-loop transfer function:

$$L(s) = \frac{1}{(0.5 + s)(s^2 + 2s + 1)},$$

Apply negative unity feedback. What can we say about the stability? Find the phase margin (pm) and the gain margin (gm), as well as the cut-off frequency wc.

```
s=zpk('s')
L=1/((0.5+s)*(s^2+2*s+1))
```

(a) Method_1: Use the margin command

```
[gm,pm,wg,wc]=margin(L)
    gm=4.5001
    pm=72.227
    wg=1.4142
    wc=0.5675
```

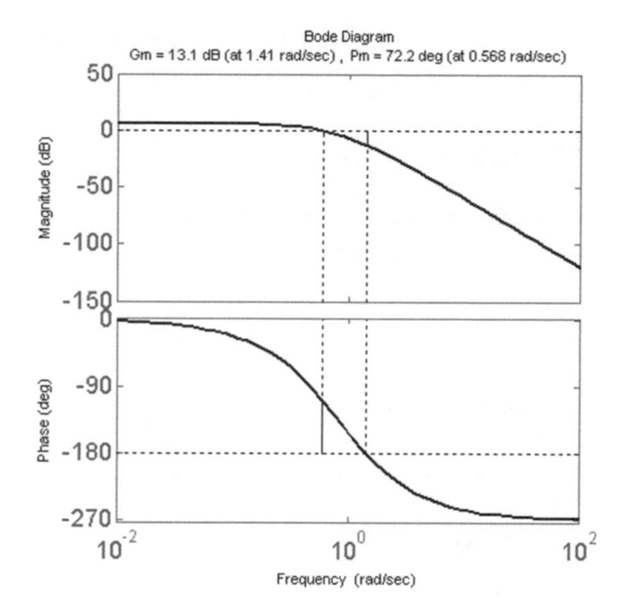

Fig. 5.9 Characteristic points of the Bode diagram

Here wg is the frequency at which the phase shift of the open-loop frequency function is $-180°$.

To get a graphical evaluation (see Fig. 5.9) we have:

```
margin(L)
```

Note that if a graphical evaluation is selected, the gain margin Gm is given in decibels.

```
Gm=20*log10(gm)
```

(b) Method_2: use the Bode and Nyquist diagrams

```
nyquist(L)
```

Read the crossing of the Nyquist plot with the negative real axis. The gain margin is

$$g_t = \mathrm{gm} = \frac{1}{|L(j\omega_\pi)|} = \frac{1}{0.22} = 4.5$$

```
bode(L)
```

Read the phase angle at the 0 dB (unity) gain (it is $-108°$). The phase margin is then obtained as $\varphi_t = \varphi(\omega_c) + 180° = -108° + 180° = 72°$.

The gain margin can be read off from the Bode diagram, just check the gain at ω_π. Clicking on the white background of the Bode diagram, select, using the right button *Properties->Units->Magnitude in—absolute*. The gain margin can be seen to be $g_t = \mathrm{gm} = \dfrac{1}{|L(j\omega_\pi)|} = \dfrac{1}{0.22} = 4.5$.

(c) Method_3: Read from a frequency-amplitude-phase table

Store the calculated points in a table then read the margins:

```
w=logspace(-1,1,100);
[num,den]=tfdata(L,'v')
[mag,phase]=bode(num,den,w);
Tabl=[mag, phase,w']
```

```
    Mag          phase           w
    1.1123      -99.5242        0.5094
    1.0643     -103.0406        0.5337
>>  1.0158     -106.6104        0.5591<<    ≈wc
    0.9669     -110.2286        0.5857

    0.2449     -176.7848        1.3530
>>  0.2211     -180.1658        1.4175<<    ≈wg
    0.1991     -183.4774        1.4850
    0.1789     -186.7160        1.5557
```

```
[mag,phase]=bode(L,w);
Tabl=[mag(:), phase(:), w']
```

LTI structure is interpreted for *MIMO* linear systems. This is why mag and phase are variables of three dimensions. In the case of *SISO* systems, the ":" operator converts the three dimensional structures to vector structures.

The phase margin can be calculated from the data in the table. To get the phase margin first the phase at the cut-off frequency should be read (the cut-off frequency is $\mathrm{w} = \mathrm{wc} = 0.5591$), then this phase value should be added to 180°: $\mathrm{pm} = 180 - 106.6 = 73.4$. The gain margin turns out to be $\mathrm{gm} = 1/0.221 = 4.52$, where 0.221 is the gain value belonging to the ω_π frequency.

Now investigate how a change in the loop gain will change the open-loop and closed-loop properties. To start, multiply the loop-transfer function by the gain margin 4.5.

```
Lk=4.5*L
nyquist(Lk)
Tk=Lk/(1+Lk)
step(Tk)
pzmap(Tk)
```

Modify the loop gain to 4.4 and then 4.6. For a loop gain like 4.6, one pole moves to the right half-plane, and the step response will diverge to infinity.

As far as the root-locus is concerned

```
rlocus(L)
```

allows sketching the critical value for the loop gain.

5.6.2 Delay Margin

The delay margin is the smallest value of the dead-time needed to push the system to borderline stability. It can be calculated as follows:

$$T_{\mathrm{min}} = \frac{\varphi_{\mathrm{t}}}{\omega_{\mathrm{c}}} = \frac{\mathsf{pm}}{\mathsf{wc}} \frac{[\mathrm{rad}]}{[\mathrm{rad/s}]}.$$

Example 5.6.2 Find the delay margin for the system discussed in Example 5.6.1.

```
Tmin=(72*pi/180)/0.56
   Tmin = 2.24
```

So a dead-time of 2.24 s can be inserted into the open-loop system to get borderline stability for the closed-loop system.

5.6.3 Modulus Margin

The modulus margin ρ_{m} is the minimum distance between the NYQUIST diagram and the point $(-1+0j)$. In other words, drawing a circle around the point $(-1+0j)$, the modulus margin will be the radius of the circle just touching the NYQUIST diagram. A practical specification for the modulus margin is $\rho_{\mathrm{m}} > 0.5$. Alternatively, the modulus margin is identical to the reciprocal of the maximum of the absolute value of the sensitivity function.

$$\rho_{\mathrm{m}} = \frac{1}{\max\limits_{\omega}|S(j\omega)|} = \min_{\omega}|1 + L(j\omega)|$$

Example 5.6.3 Find the modulus margin of the system discussed in Example 5.6.1.

```
M=bode(L+1)
ro=min(M)
   ro = 0.6317
```

5.7 Robust Stability

A closed-loop is robustly stable if it remains stable in spite of uncertainities in the process to be controlled.

For a process with nominal model $\hat{P}$, suppose that its real transfer function is P. Then the relative uncertainty is $\ell = \frac{P-\hat{P}}{\hat{P}}$. Robust closed-loop stability is achieved if $|\ell(j\omega)| < \frac{1}{|\widehat{T}_{\rm m}|}$ holds for all frequencies, where $\widehat{T}_{\rm m}$ is the maximum value of the complementary sensitivity function of the nominal closed-loop system. Note that the complementary sensitivity function equals the overall transfer function of the closed-loop system between the process output and the reference input. (See Eq. 5.44 in the textbook [1].)

Example 5.7.1 Consider the transfer function of a nominal process: $\hat{P}(s) = \frac{1}{(1+s)(1+5s)(1+10s)}$.

The time constants and the gain of the real process, however differ from those of the nominal transfer function:

$$P(s) = \frac{1.2}{(1+2s)(1+6s)(1+12s)}.$$

The following series regulator has been designed for the nominal process:

$$C(s) = 2.5\frac{1+10s}{10s}\frac{1+5s}{1+s}.$$

Check the stability of the closed-loop system containing the real process driven by the series regulator designed for the nominal process. The open-loop transfer function with the nominal process is determined by

$$L(s) = C(s)\hat{P}(s) = \frac{0.25}{s(1+s)^2}.$$

Check whether the condition of robust stability is satisfied:

```
s=zpk('s')
Pk=1/((1+s)*(1+5*s)*(1+10*s))
P=1.2/((1+2*s)*(1+6*s)*(1+12*s))
L=0.25/(s*(1+s)*(1+s))
T=L/(1+L)
T=minreal(T)
l=(P-Pk)/Pk
l=minreal(l)
w=logspace(-2,1,200);
[magT,phaseT]=bode(T,w);
```

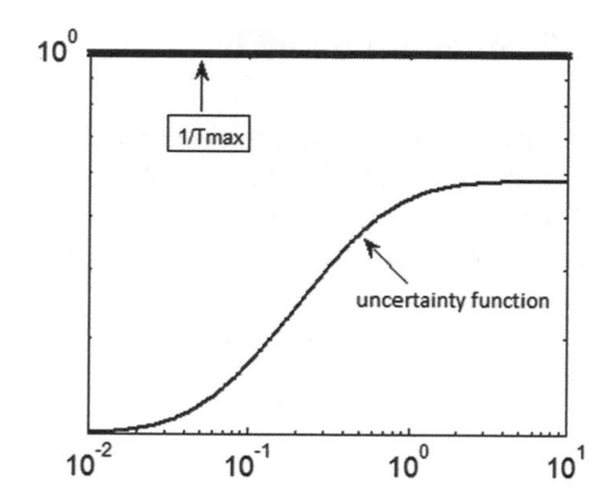

Fig. 5.10 Uncertainty function

```
Tmax=max(magT(:))
[magl,phasel]=bode(1,w);
loglog(w,magl(:),'r',w,1/Tmax,'bx')
```

Figure 5.10 shows that the frequency function of the relative uncertainty runs below the value of 1/Tmax in the whole frequency range, so the condition for robust stability holds. The step responses of the nominal (solid line) and real (dotted line) system (Fig. 5.11) differ. The transient of the real system exhibits a higher overshoot, but it remains stable.

The MATLAB™ commands to calculate the step responses are:

```
C=2.5*(1+10*s)*(1+5s)/((10*s)*(1+s))
L1=C*P
T1=L1/(1+L1)
t=0:0.1:40;
y=step(T,t);
y1=step(T1,t);
plot(t,y,t,y1), grid
```

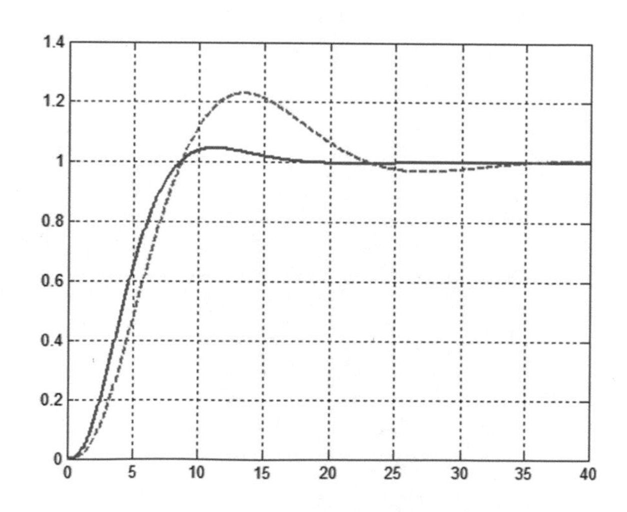

Fig. 5.11 Step responses of the nominal and the real closed-loop system

5.8 Internal Stability

A closed-loop system is internally stable (in the *BIBO* sense) if upon applying any bounded external excitation, all the internal signals in the system remain bounded. The reference signal, the disturbance acting at the input and the disturbance at the output of the process and the measurement noise signal are considered as external excitations.

For internal stability, all the entries in the transfer function matrix T_t must be stable:

$$T_t = \begin{bmatrix} \frac{CP}{1+CP} & \frac{P}{1+CP} \\ \frac{C}{1+CP} & \frac{1}{1+CP} \end{bmatrix}$$

Example 5.8.1 Consider the unstable process $P(s) = \dfrac{10}{s-1}$

Can we stabilize (in the internal sense) the closed-loop system using a regulator given by $C(s) = \dfrac{s-1}{s}$?

Find the transfer functions involved by using the T_t matrix:

```
s=zpk('s')
P=10/(s-1)
C=(s-1)/s
L=C*P, L=minreal(L)
T11=L/(1+L); T11=minreal(T11)
T12=P/(1+L); T12=minreal(T12)
T21=C/(1+L); T21=minreal(T21)
T22=1/(1+L); T22=minreal(T22)
```

The results give:

```
  10
------;
(s+10)

     10 s
------------;
(s-1)(s+10)

 (s-1)
------;
(s+10)

   s
------.
(s+10)
```

It can be seen that one of the transfer functions (T12) has a pole in the right half-plane, thus the system is internally unstable.

Note that direct cancellation of an unstable pole is not an accepted design procedure, as it results in unstable internal behaviour.

Example 5.8.2 Recap the design problem introduced in the previous example: $P(s) = \frac{10}{s-1}$.

Can we stabilize (in the internal sense) the closed-loop system using a regulator by $C(s) = \frac{s+1}{s}$?

Find the transfer functions involved by using the T_t matrix.

```
s=zpk('s')
P=10/(s-1)
C=(s+1)/s
L=C*P;L=minreal(L)
T11=L/(1+L);T11=minreal(T11)
T12=P/(1+L);T12=minreal(T12)
T21=C/(1+L);T21=minreal(T21)
T22=1/(1+L);T22=minreal(T22)
```

The elements of the transfer function matrix are:

```
        10 (s+1)
-------------------;
(s+1.298) (s+7.702)

        10 s
-------------------;
(s+1.298) (s+7.702)

    (s+1)(s-1)
-------------------;
(s+1.298) (s+7.702)

      s (s-1)
-------------------.
(s+1.298) (s+7.702)
```

The poles of each transfer function have negative real parts, so the closed-loop system has become internally stable.

Chapter 6
Design in the Frequency Domain

A closed loop control system has to meet several prescribed quality specifications. These specifications can be formulated in the time domain and also in the frequency domain. The behaviour of the closed loop control system can be evaluated on the basis of the frequency function of the open loop. The characteristics of the open loop frequency function in the low-, middle- and high frequency ranges determine the quality characteristics of the closed loop system.

- For good reference signal tracking, $|L(j\omega)|$ should be large in the low frequency range.
- For effective rejection of measurement noise, $|L(j\omega)|$ should be small in the high frequency range.
- For faster performance, the cut-off frequency ω_c should be as large as possible.
- To ensure stability, the cut-off frequency should be located at that part of the Bode amplitude diagram where the slope of its asymptote is −20 dB/decade.
- The overshoot of the step response will be within 10%, if the phase margin is about 60°.

These requirements are partly contradictory. Different prescriptions have to be given for different frequency ranges. The requirements can be fulfilled by appropriate shaping of the frequency characteristics. Figure 6.1 shows a typical open loop amplitude-frequency function.

Example 6.1 Let us consider a system whose loop frequency function shows similar performance to Fig. 6.1. Analyse the static response, reference signal tracking, disturbance rejection and transient behaviour of the closed loop system.

$$L(s) = C(s)P(s) = 0.01(20s+1)/\left[s^2(s+1)\right]$$

L. Keviczky et al., *Control Engineering: MATLAB Exercises*,
Advanced Textbooks in Control and Signal Processing,
https://doi.org/10.1007/978-981-10-8321-1_6

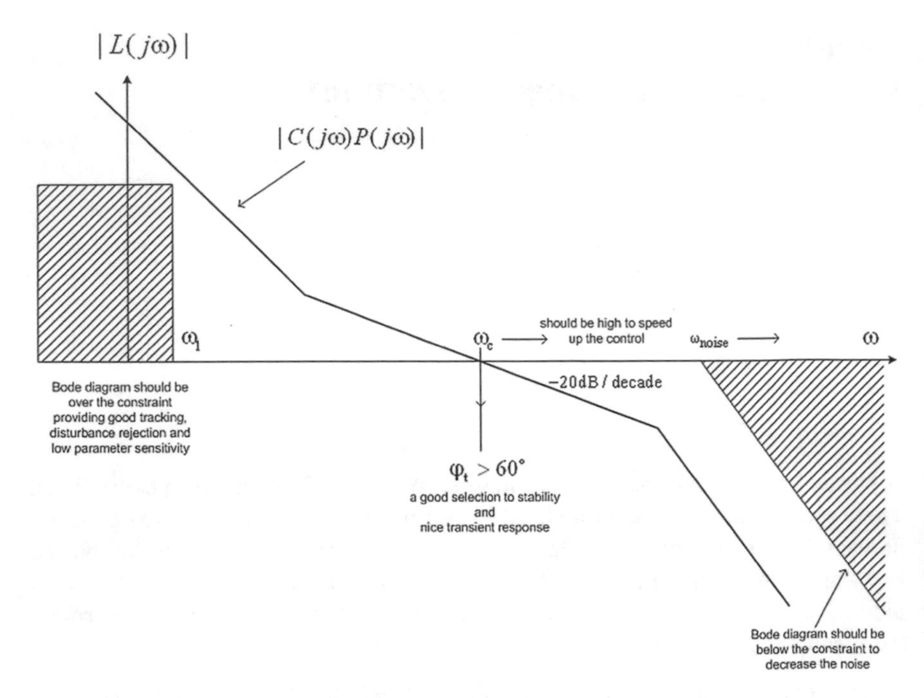

Fig. 6.1 Formulation of the quality specifications in the frequency domain

Plot the Bode diagram of the open loop.

```
s=zpk('s')
L=0.01*(20*s+1)/(s*s*(s+1))
T=L/(1+L),T = minreal(T)
figure(1),bode(L),grid
```

The Bode diagram is shown in Fig. 6.2. Analyse the behaviour of the system for reference signal tracking and disturbance rejection. Let us take a reference signal containing two components: a low and a high frequency sinusoidal signal.

$$r(t) = r_a(t) + r_m(t) = \sin(\omega_a t) + 0.5\sin(\omega_m t)$$

Let ω_a be a low frequency which is located in the first part of the Bode diagram of slope −40 dB/decade, and ω_m be on the second part on the high frequency part with the same slope: $\omega_a = 0.05$, $\omega_m = 2$. Plot the input and the output signal!

```
t=0:0.1:500;
wa=0.05; wm=2;
r=sin(wa*t)+0.5*sin(wm*t);
y=lsim(T,r,t);
figure(2);
subplot(211); plot(t,r);
subplot(212); plot(t,y);
```

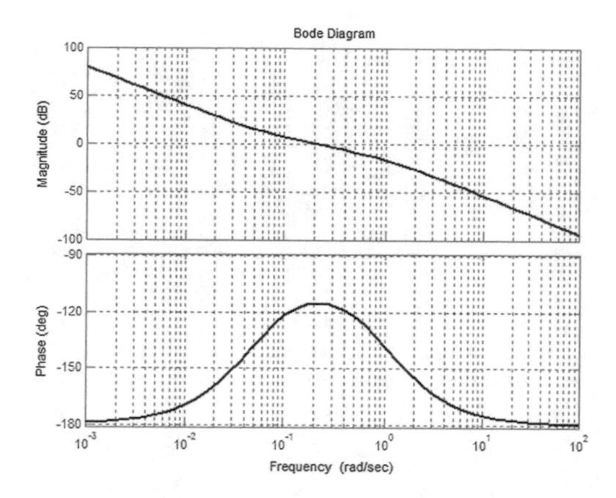

Fig. 6.2 Open-loop frequency function

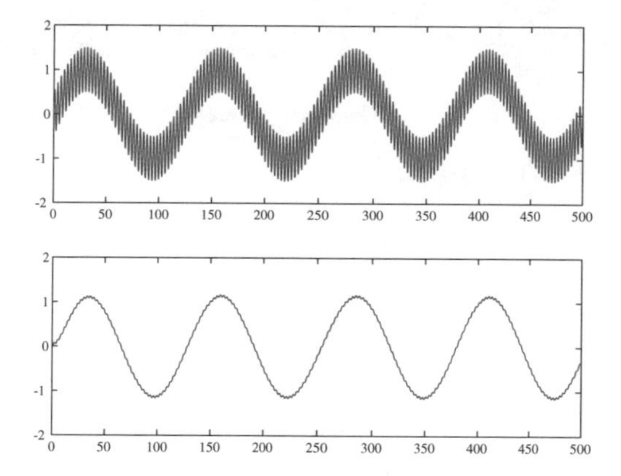

Fig. 6.3 The system attenuates the high frequency component

In Fig. 6.3 it can be seen that the system with such frequency characteristics tracks the low frequency component of the input signal and attenuates the high frequency component (the upper curve in the figure is the input signal, and the lower curve is the output signal).

The transient behaviour can be analysed on the step response of the closed loop system.

```
figure(3),step(T),grid
```

The overshoot depends on the phase margin, which can read from the Bode diagram ($\sim 60°$). The settling time depends on the cut-off frequency ω_c.

Example 6.2 Let us analyse the relationship between the cut-off frequency and the settling time. Choose now a faster L_1 system than the system L in Example 6.1. The cut-off frequency of system L_1 is higher than that of system L.

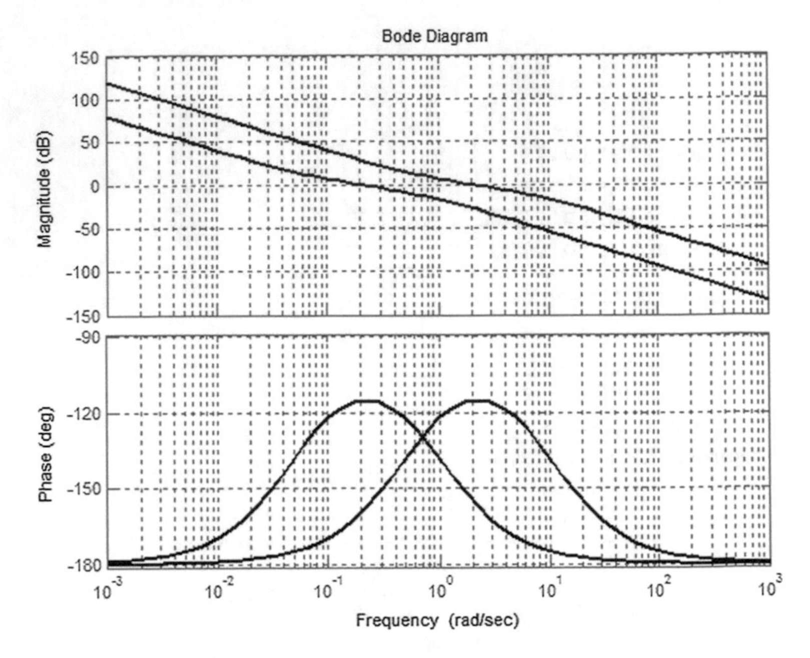

Fig. 6.4 Bode diagrams of two systems

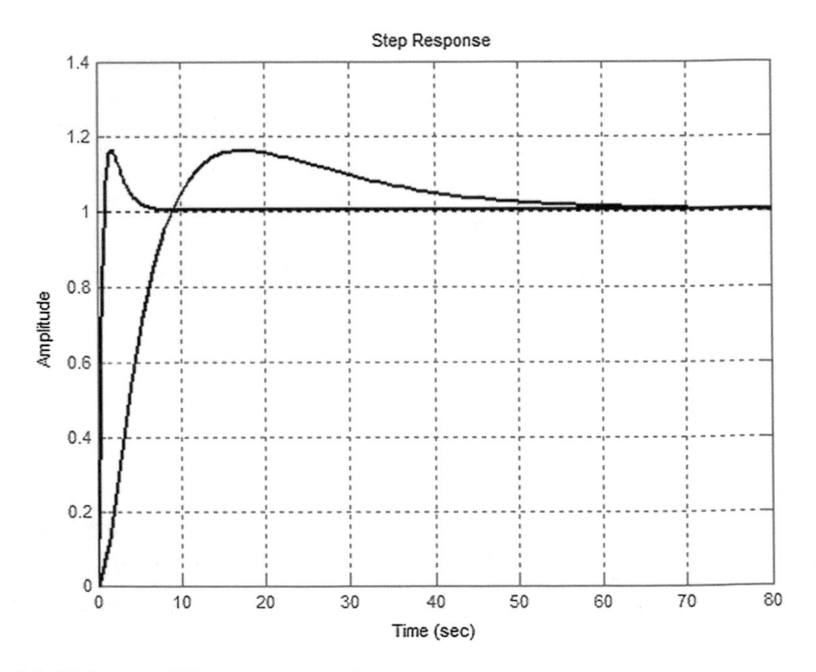

Fig. 6.5 Higher cut-off frequency means faster step response

$$L_1(s) = \frac{2s+1}{s^2(0.1s+1)}$$

```
L1=(2*s+1)/(s*s*(0.1*s+1))
T1=L1/(1+L1),T1=minreal(T1)
figure(1),bode(L,'b',L1,'r'),grid
figure(3),step(T,'b',T1,'r'),grid
```

It can be seen (Figs. 6.4 and 6.5) that higher cut-off frequency means a faster time response.

The cut-off frequency of the first system is $\omega_c \approx 0.2$ and that of the second system is $\omega_c \approx 2$. The settling time for the first system is $t_s \approx 50$ s, while for the second system it is $t_s \approx 5$ s (The settling time can be estimated by the relation $3/\omega_c < t_s < 10/\omega_c$).

As the open loop contains two integrators, the control system tracks the step and also the ramp input accurately, without steady error. Show this by simulation!

Chapter 7
Control of Stable Continuous Processes, Youla Parameterization

Youla parameterization can be used to control stable processes. The block diagram of a common control system is shown in Fig. 7.1. With the appropriate regulator design, a closed loop control based on negative feedback ensures reference signal tracking, rejection of the effects of input and output disturbances, and also attenuation of the measurement noise.

Denote by Q the resulting transfer function between the control signal (manipulated variable) u and the reference signal r.

$$\frac{U(s)}{R(s)} = \frac{C}{1 + CP} = Q$$

The resulting transfer function between the output signal y and the reference signal r, supposing that the other input signals are zero, is $T = QP$. Therefore if P is stable, any stable Q ensures a stable closed loop control circuit.

The parameter Q (which is a transfer function) is called the Youla parameter.

The series regulator $C(s)$ can be expressed also using the $Q(s)$ parameter:

$$C(s) = \frac{Q}{1 - QP}$$

Let us remark that the resulting transfer function $T = QP$ is linear in the parameter Q, whereas in the regulator C, it is nonlinear. The aim of regulator design is to fulfill the prescribed quality specifications set for the control system. Because of the linearity of the relation, the design of Q is simpler than the design of C.

In a closed loop control system the resulting transfer functions can be expressed also by the Youla parameter Q.

The servo (tracking) property (reference signal => output signal): $T = QP$ (this is the complementary sensitivity function)

Disturbance rejection (output disturbance => output signal, or reference signal => error signal):

L. Keviczky et al., *Control Engineering: MATLAB Exercises*,
Advanced Textbooks in Control and Signal Processing,
https://doi.org/10.1007/978-981-10-8321-1_7

$$S = \frac{1}{1+CP} = \frac{1+CP-CP}{1+CP} = 1 - \frac{CP}{1+CP} = 1 - QP \text{ (sensitivity function)}$$

and $T + S = 1$ is fulfilled.

Relations between the output and the input signals:

The input signals are the reference signal and the output disturbance: r, $y_n = y_{no}$. The output signals are the control signal, the error signal and the output signal: u, e, y.

$$\begin{bmatrix} u \\ e \\ y \end{bmatrix} = \begin{bmatrix} Q & -Q \\ 1-QP & -(1-QP) \\ QP & 1-QP \end{bmatrix} \begin{bmatrix} r \\ y_n \end{bmatrix} = \begin{bmatrix} Q & -Q \\ S & -S \\ T & S \end{bmatrix} \begin{bmatrix} r \\ y_n \end{bmatrix}$$

The aim is always to ensure $T \Rightarrow 1$ (good servo property) over a wide frequency range, and $S \Rightarrow 0$ (good disturbance rejection). In the ideal case $T = 1$ and $S = 0$.

$$y = Tr + Sy_n \Rightarrow r \quad \text{or} \quad e = Sr - Sy_n \Rightarrow 0.$$

Regarding reference signal tracking Fig. 7.2 is equivalent to Fig. 7.1.

The best reference signal tracking is obtained if Q is the inverse of the transfer function of the process, i.e. $Q = P^{-1}$.

But this open loop control structure does not ensure disturbance rejection. Supplement this circuit with the inner model according to Fig. 7.3 (here only the output disturbance is shown). This is the so called *Internal Model Control* (*IMC*) structure.

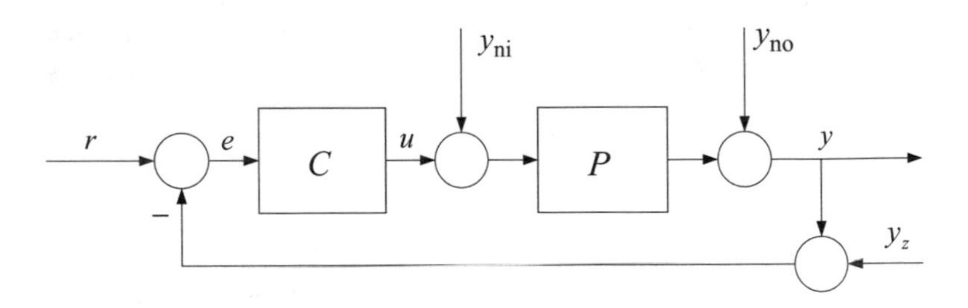

Fig. 7.1 Block diagram of a control system

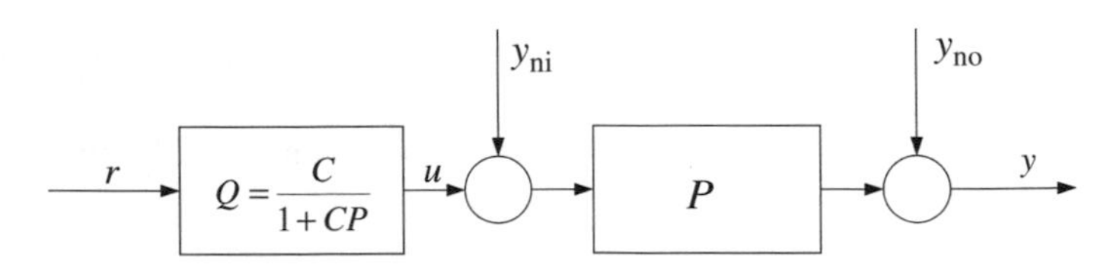

Fig. 7.2 Demonstration of the YOULA parameter

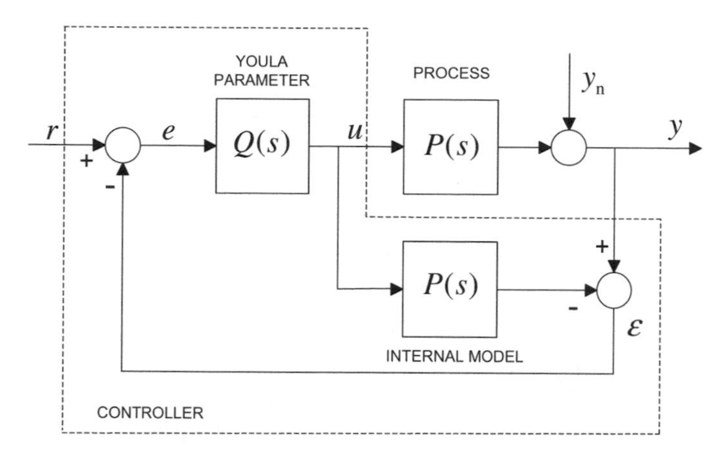

Fig. 7.3 Supplementing the control circuit with internal model control

If there is no disturbance and the internal model is exactly the same as the process, then the value of the feedback signal is zero, and reference signal tracking is determined by the forward path of the open loop. Feedback will ensure disturbance rejection and eliminating the effect of plant-model mismatch.

If the model is perfect, then $\varepsilon = y_n$, and then the input of element Q is $r - \varepsilon = r - y_n$. Thus

$$u = Q(r - y_n)$$
$$y = Pu + y_n = PQr - PQy_n + y_n = Tr - Ty_n + y_n = Tr + (1 - T)y_n.$$

The block diagram in Fig. 7.3 can be redrawn according to Fig. 7.4.

Figure 7.4 is equivalent to the usual feedback control system shown in Fig. 7.5 and also in Fig. 7.1.

The series regulator $C(s)$ can be expressed with the YOULA parameter $Q(s)$, as was shown previously:

$$C(s) = \frac{Q}{1 - QP}.$$

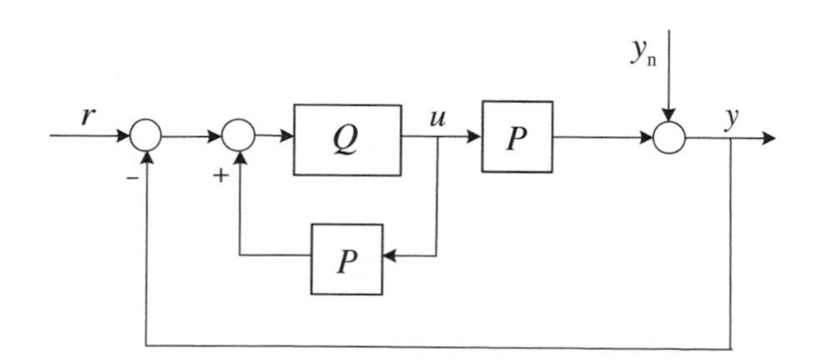

Fig. 7.4 Equivalent control scheme

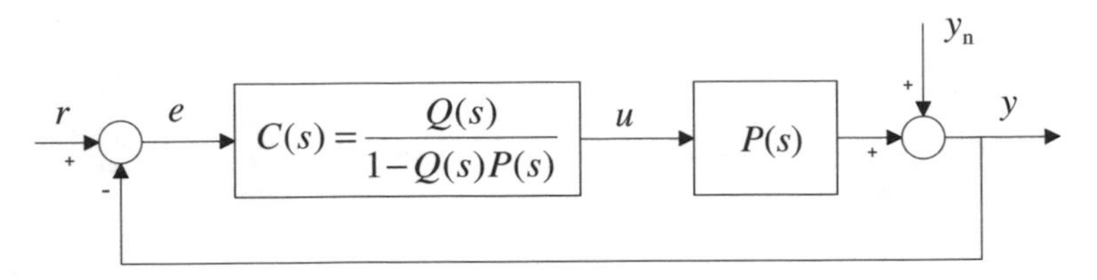

Fig. 7.5 Equivalent usual feedback control scheme

The two structures (Figs. 7.3 and 7.5)—as a *YP* regulator—give results equivalent to the closed loop circuit.

Generally ideal reference signal tracking can not be realized, since the transfer function of the process can not always be inverted. Dead-time is non-invertible, as its inverse is not realizable. A transfer function can not be inverted if the degree of its denominator polynomial is higher than the degree of its numerator polynomial, as in this case the inverse is not realizable. A non-minimum phase transfer function, which contains zeros in the right half-plane is not invertible either, as its inversion would introduce unstable poles in the transfer function of the regulator.

Let us express $P(s)$, the transfer function of the process, as the product of its invertible part $P_+(s)$ and non-invertible $\bar{P}_-(s)$ part. $\bar{P}_-(s) = P_-(s)e^{-sT_d}$ contains the non-invertible part of the process transfer function and the dead-time. The gain of $P_-(s)$ should be 1 to ensure accurate reference tracking in the control system: $P_-(s=0) = 1$.

$$P(s) = P_+(s)\bar{P}_-(s)$$

When realizing the control system the YOULA parameter performs the inverse of the invertible part only (Fig. 7.6): $Q = P_+^{-1}$.

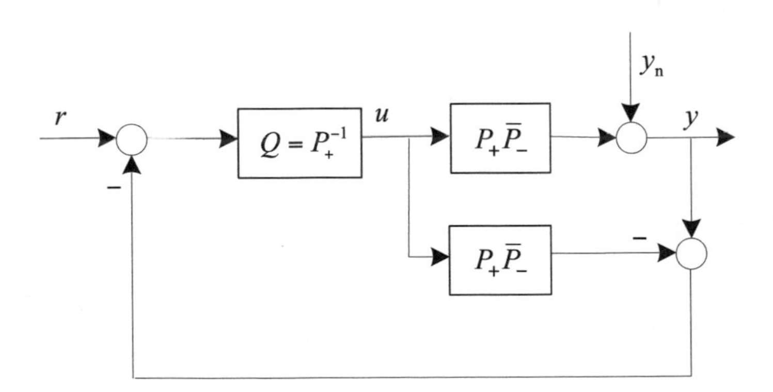

Fig. 7.6 The YOULA parameter is the inverse of the invertible part of the process

The dynamics of reference signal tracking and disturbance rejection are the same in this structure.

$$\left.\frac{Y(s)}{R(s)}\right|_{y_n=0} = \overline{P}_-(s) \quad \text{and} \quad \left.\frac{Y(s)}{Y_n(s)}\right|_{r=0} = 1 - \overline{P}_-(s)$$

It is often required that the dynamics of tracking be different than the dynamics of disturbance rejection, for example, that the disturbance rejection should be faster than the reference signal tracking. This is expressed also as that the *one-degree-of-freedom* (*1DOF*) control should be converted to *two-degree-of-freedom* (*2DOF*) control.

This can be ensured by using filters $R_r(s)$ and $R_n(s)$ according to Fig. 7.7. The gain of the filters should be unity to ensure the correct static values.

In this structure, the dynamics of the reference signal tracking and that of the disturbance rejection are obtained by the following:

$$\left.\frac{Y(s)}{R(s)}\right|_{y_n=0} = R_r(s)\overline{P}_-(s) \quad \text{and} \quad \left.\frac{Y(s)}{Y_n(s)}\right|_{r=0} = 1 - R_n(s)\overline{P}_-(s).$$

A further role of the filters is that they modify the maximum value of the control signal u (the manipulated variable), so it can be ensured that it will not exceed its limit. The filters may have also a robustification effect. With their appropriate choice, the control system could be made less sensitive to plant-model mismatch.

Figure 7.7 can be redrawn according to Fig. 7.8. The YOULA parameter now is $Q = R_n P_+^{-1}$.

With further restructuring, a conventional feedback structure is obtained (Fig. 7.9), where the series regulator is expressed in the form: $C = \frac{R_n P_+^{-1}}{1 - R_n \overline{P}_-} = \frac{Q}{1 - QP}$.

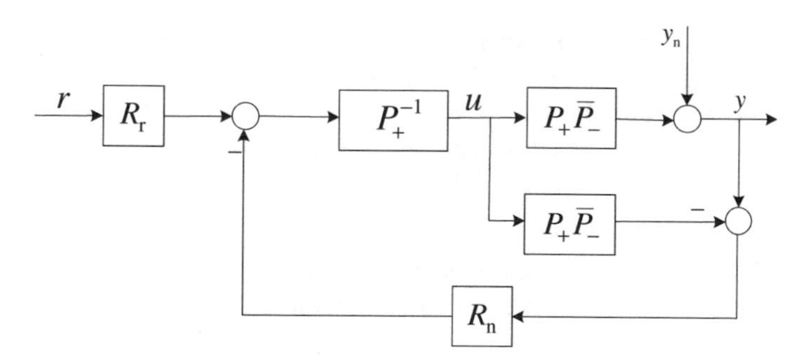

Fig. 7.7 Introducing filters in the control scheme

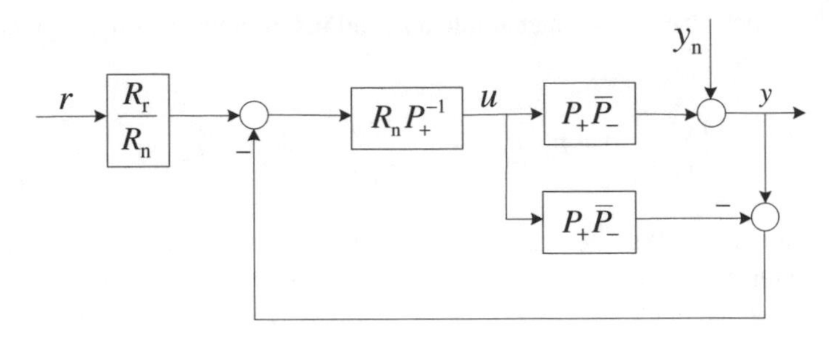

Fig. 7.8 Equivalent control scheme

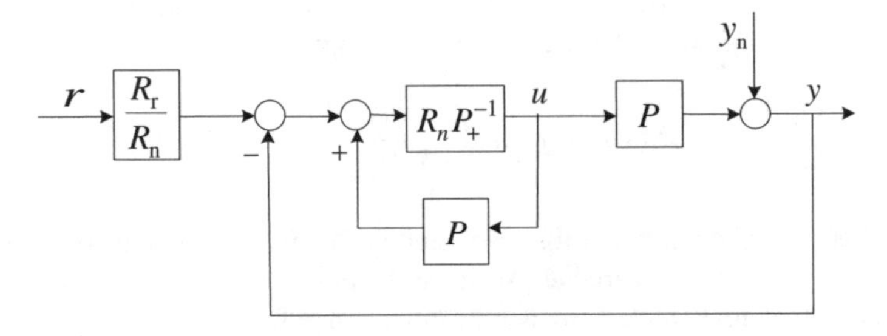

Fig. 7.9 Equivalent control scheme

With these filters, the relations between the signals can be expressed as

$$\begin{bmatrix} u \\ e \\ y \end{bmatrix} = \begin{bmatrix} \frac{R_r}{R_n} Q & -Q \\ \frac{R_r}{R_n}(1-QP) & -(1-QP) \\ \frac{R_r}{R_n} QP & 1-QP \end{bmatrix} \begin{bmatrix} r \\ y_n \end{bmatrix} = \begin{bmatrix} \frac{R_r}{R_n} Q & -Q \\ \frac{R_r}{R_n} S & -S \\ \frac{R_r}{R_n} T & S \end{bmatrix} \begin{bmatrix} r \\ y_n \end{bmatrix}$$

Summarizing: YOULA parameterization gives a method for regulator design (determines the Q parameter in IMC structure, or regulator C in feedback structure). For the design the transfer function of the process has to be separated into the invertible and non-invertible components and furthermore the transfer functions of the reference and disturbance filters have to be given.

Examples First let us consider processes without dead-time.

Write a MATLAB™ program, where the inputs are P_+ and P_-, the invertible and non-invertible components of the process, respectively, and the filters R_r and

R_n. The program calculates and plots the output and control signals for unit step reference signal and zero output disturbance, and then the output and the control signals for zero reference signal and unit step output disturbance.

Let us save the program with the name Youla_cont.

```
% Youla_cont: Youla continuous basic program
display('..... Q='),Q=minreal(Rn/Pp,0.0001)
display('..... C='),C=minreal(Q/(1-Q*P),0.0001)
display('..... Tr='),Tr=minreal((Rr/Rn)*Q*P,0.0001)
display('..... Ur='),Ur=minreal((Rr/Rn)*Q,0.0001)
pause;
t=0:0.1:50;
figure(1)
y=step(Tr,t); subplot(211),plot(t,y),grid;
u=step(Ur,t); subplot(212),plot(t,u),grid;
pause;
display('.....Sn='),Sn=minreal((1-Q*P),0.0001);
display('.....Un'),Un=-Q;
figure(2)
y=step(Sn,t); subplot(211),plot(t,y),grid;
u=step(Un,t); subplot(212),plot(t,u),grid;
```

Example 7.1 Consider *Example 7.2* of the textbook [1].

The process to be controlled is a second-order proportional system given by the transfer function

$$P = \frac{(1+5s)(1+6s)}{(1+10s)(1+8s)} = P_+$$

This system is invertible, so $P_- = 1$.

The filters are $R_r = \frac{1}{1+4s}$ and $R_n = \frac{1}{1+2s}$.

Give the data in MATLAB™.

```
s=zpk('s')
P=((1+5*s)*(1+6*s))/((1+10*s)*(1+8*s))
Pp=P
Rr=1/(1+4*s)
Rn=1/(1+2*s)
```

Then call the program Youla_cont.

```
Youla_cont
```

The calculated Q and C are

```
        1.3333(s+0.1)(s+0.125)
Q=  -------------------------
       (s+0.5)(s+0.2)(s+0.1667)

        1.3333(s+0.1)(s+0.125)
C=  -----------------------
          s(s+0.2)(s+0.1667)
```

It can be seen that the regulator C contains an integrator, so the control system will track the reference signal without steady error.

The left side of Fig. 7.10 gives the output and the control signals for reference signal tracking, while the right side of the figure shows these signals for the output disturbance. It is seen that the disturbance rejection is faster than the reference signal tracking.

Example 7.2 The process is of non-minimum phase with one zero in the right half-plane. Its transfer function is

$$P = \frac{1-s}{(1+s)(1+2s)}$$

Its step response is shown in Fig. 7.11. It can be seen that the response starts downwards, then turns and reaches the steady state value determined by unit gain.

Let us separate the transfer function of the process into its invertible and non-invertible components: $P_- = 1 - s$ (its gain, which is its value at, $s = 0$ is 1)

$$P_+ = \frac{1}{(1+s)(1+2s)}$$

The filters have to be of second-order to ensure the realizability of Q and C, and also of ratio R_r/R_n.

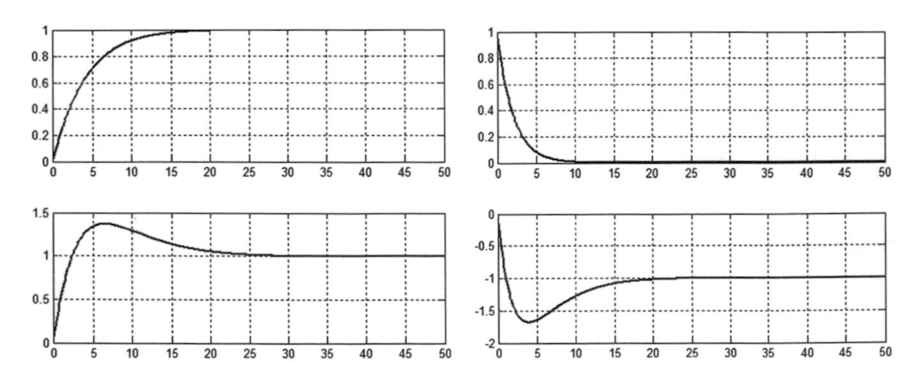

Fig. 7.10 Output and control signals

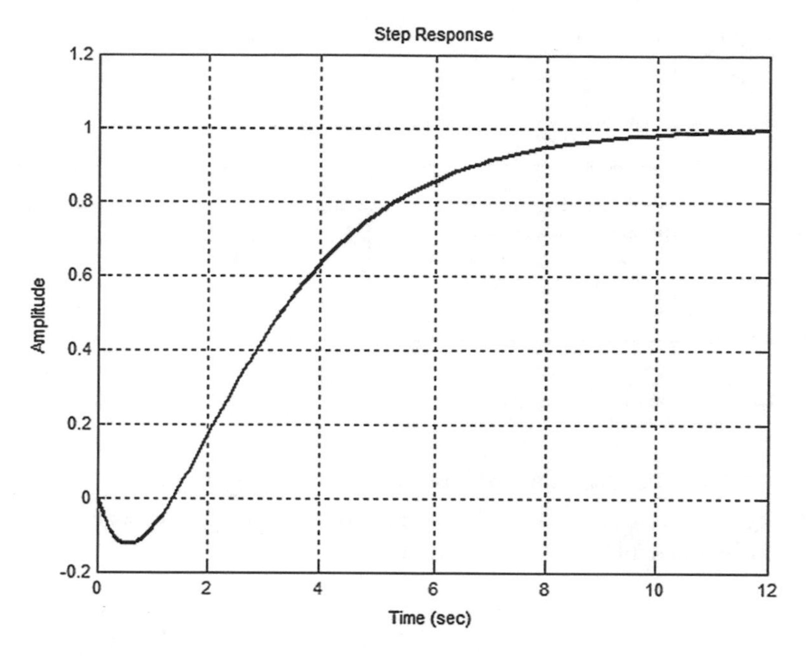

Fig. 7.11 Step response of a non-minimumphase process

Put $R_\mathrm{r} = \frac{1}{(1+s)^2}$ and $R_\mathrm{n} = \frac{1}{(1+0.5s)(1+s)}$.

Give the data in MATLAB™.

```
s=zpk('s')
P=(1-s)/((1+s)*(1+2*s))
Pp=1/((1+s)*(1+2*s))
Rr=1/(1+s)^2
Rn=1/((1+0.5*s)*(1+s))
```

Then call the Youla_cont program.

```
Youla_cont
```

It can be seen that the control system operates appropriately (Fig. 7.12). The left side of the figure gives the output and the control signals for reference signal tracking, while the right side shows these signals for the output disturbance. (Without the separation of the transfer function of the process with the original algorithm the control has unstable behaviour.) The settling time of the control system is about the half of that of the process itself.

Example 7.3 Let us consider now a process with dead-time, where the value of the dead-time is bigger than the values of the time constants in the lag elements. In such cases a YOULA parameterized regulator results in a significantly faster control system than *PID* control (see Chap. 8).

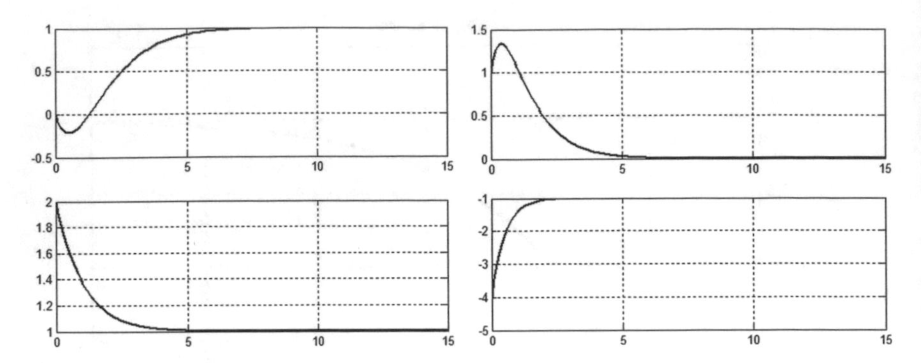

Fig. 7.12 Output and control signals

The transfer function of the process is: $P(s) = \frac{1}{(1+5s)(1+10s)} e^{-30s}$.

Design a YOULA parameterized regulator for this process. Separate the transfer function into invertible and non-invertible parts:

$$P_{+} = \frac{1}{(1+5s)(1+10s)} \quad \text{and} \quad \bar{P}_{-} = P_{-}e^{-30s} = 1 \cdot e^{-30s}$$

The filters have to be of second order to ensure realizable Q and $R_{\mathrm{r}}/R_{\mathrm{n}}$ transfer functions.

Let us first choose two identical filters, $R_{\mathrm{r}} = R_{\mathrm{n}} = \frac{1}{(1+s)^2}$, then put $R_{\mathrm{r}} = \frac{1}{(1+s)^2}$ and $R_{\mathrm{n}} = \frac{1}{(1+25s)(1+s)}$.

Analyse the reference signal tracking and disturbance rejection of the control system for unit step reference signal and disturbance, first when the system and its model are the same, then when there is plant-model mismatch: the dead-time of the process is 40 s, and the dead-time in the model, which is the basis of regulator design is 30 s.

The simulation is executed in SIMULINK™, as in MATLAB™ the dead-time could be considered only approximately.

Give the data in MATLAB™ and the commands calculating the YOULA parameter and the filters.

```
s=zpk('s')
Pp=1/((1+5*s)*(1+10*s))
Rr1=1/(1+s)^2
Rn1=1/(1+s)^2
Rr2=Rr1;
Rn2=1/((1+25*s)*(1+s))
Q1=minreal(Rn1/Pp,0.0001)
Q2=minreal(Rn2/Pp,0.0001)
F1=1;
```

```
F2=minreal(Rr2/Rn2,0.0001)
```

Build the SIMULINK™ block diagram according to the *IMC* structure (Fig. 7.13).

SIMULINK™ takes over the referred variables given in MATLAB™.

The reference signal is a unit step acting at $t = 0$, the output disturbance is a step with amplitude 0.5 which acts at time $t = 100$. Figure 7.14 shows the output signals of the two filter choices. The control signals could be also plotted.

Figure 7.15 shows the output signals in the case of plant-model mismatch. With the filters of smaller time constants the control system becomes unstable, but with the bigger time constants in the filters the control remains stable.

Let us remark that in case of a sampled-data system, the realization of a regulator containing dead-time does not present a problem, and so the simulation can be executed easily also in MATLAB™ (see Sect. 12.1).

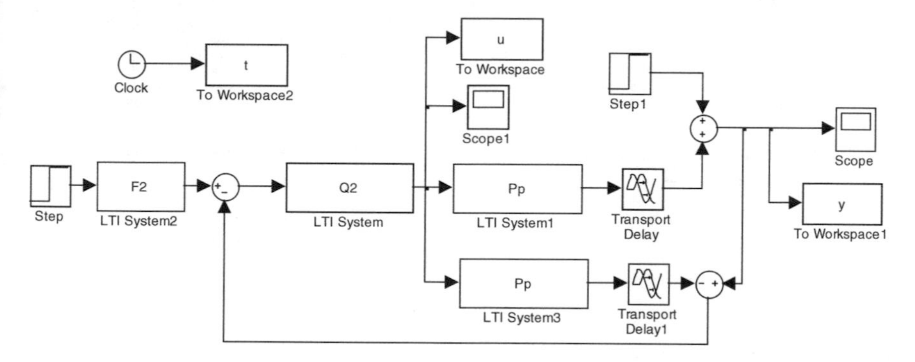

Fig. 7.13 SIMULINK™ block diagram of the Youla parameterized control system

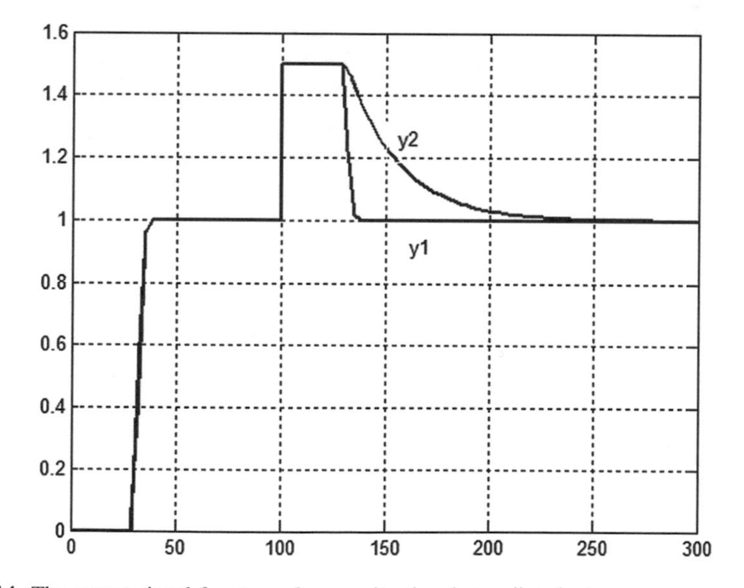

Fig. 7.14 The output signal for step reference signal and step disturbance

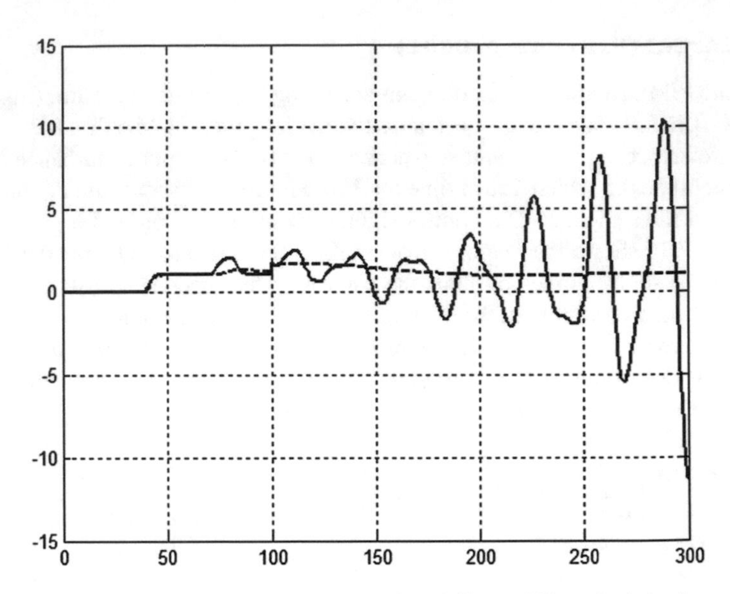

Fig. 7.15 In case of dead time mismatch a well selected filter stabilizes the behavior

Chapter 8
PID Regulator Design

8.1 Characteristics of *PID* Elements

8.1.1 Characteristics of the PI Element

The transfer function of an ideal *PI* element is $C(s) = A_\mathrm{P}\frac{1+sT_\mathrm{I}}{sT_\mathrm{I}} = A_\mathrm{P}\left(1 + \frac{1}{sT_\mathrm{I}}\right)$. Plot its step response in case of $T_\mathrm{I} = 10$ and $A_\mathrm{P} = 2$.

```
s=tf('s')
C=2*(1+s*10)/(s*10);
step(C),grid
```

It can be seen that the initial jump of the curve is $A_\mathrm{P} = 2$ (which can be calculated by limit values for $t \to 0$ or $s \to \infty$). The slope of the curve depends on the time constant $T_\mathrm{I} = 10$ (Fig. 8.1).

Let us draw the Bode diagram of the element (Fig. 8.2).

```
bode(C),grid
```

In the low frequency range the Bode amplitude diagram can be approximated by a straight line of slope −20 dB/decade, then from frequency $1/T_\mathrm{I} = 0.1$ with a horizontal straight line (Fig. 8.2).

L. Keviczky et al., *Control Engineering: MATLAB Exercises*,
Advanced Textbooks in Control and Signal Processing,
https://doi.org/10.1007/978-981-10-8321-1_8

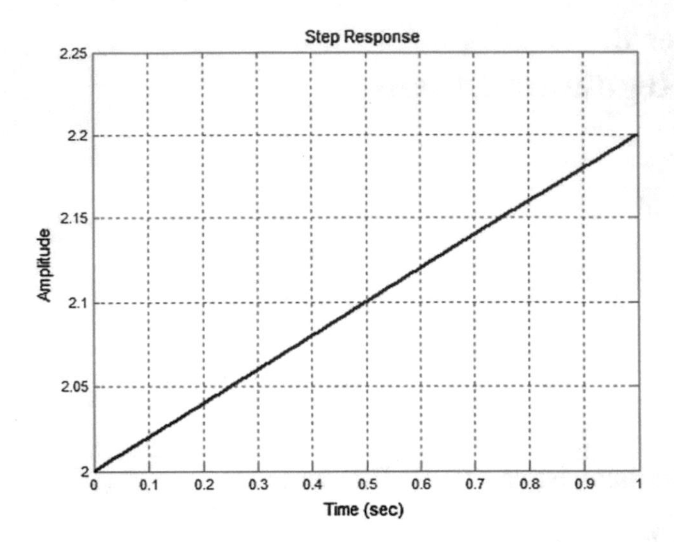

Fig. 8.1 Step response of a *PI* element

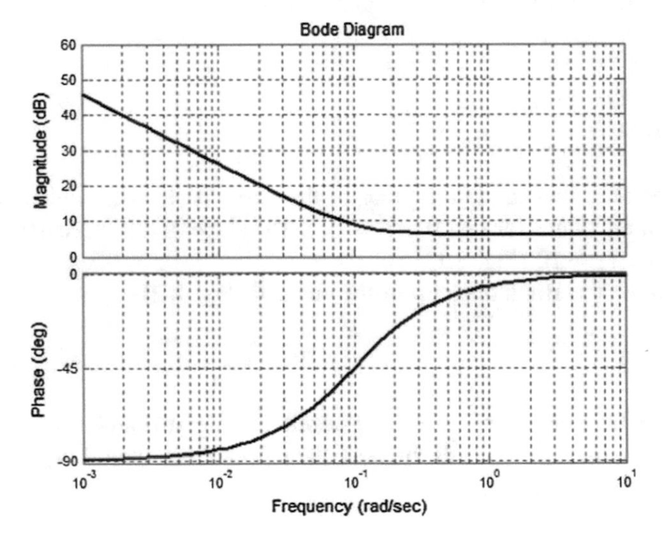

Fig. 8.2 Bode diagram of a *PI* element

8.1.2 Characteristics of the PD Element

The transfer function of the ideal *PD* element is $C(s) = 1 + sT_D$. Plot its step response in case of $T_D = 10$;

```
s=tf('s')
C=1+s*10;
step(C)
```

MATLAB™ respond with an error message, as the regulator is non-realizable: the degree of its numerator is higher than the degree of its denominator. Its step response is DIRAC delta, which can not be handled numerically.

```
??? Error using ==> rfinputs
Not supported for non-proper models.
```

The transfer function of a non ideal *PD* element is

$$C(s) = A_P\left(1 + \frac{s\tau}{1+sT}\right) = A_P\left(\frac{1+sT_D}{1+sT}\right), \; T_D = T + \tau > T.$$

Plot its step response and BODE diagram.

```
C=(1+s*10)/(1+2*s)
figure(1),step(C)
figure(2),bode(C)
```

It can be seen that the step response in point $t = 0$ starts at 5 and in steady state approximates 1 (Fig. 8.3). The ratio of the initial and final values gives the overexcitation, which ensures acceleration in the control system. The BODE diagram is shown in Fig. 8.4. Between the frequencies 0.1 and 0.5 the amplitude diagram can be approximated by a straight line of slope +20 dB/decade. This regulator is also called phase lead regulator as its phase angle is positive over the whole frequency range.

8.1.3 Characteristics of the PID Element

A *PID* regulator can be built of a proportional element (*P*), an integrating element (*I*) and a differentiating element (*D*). For regulator design in the frequency domain

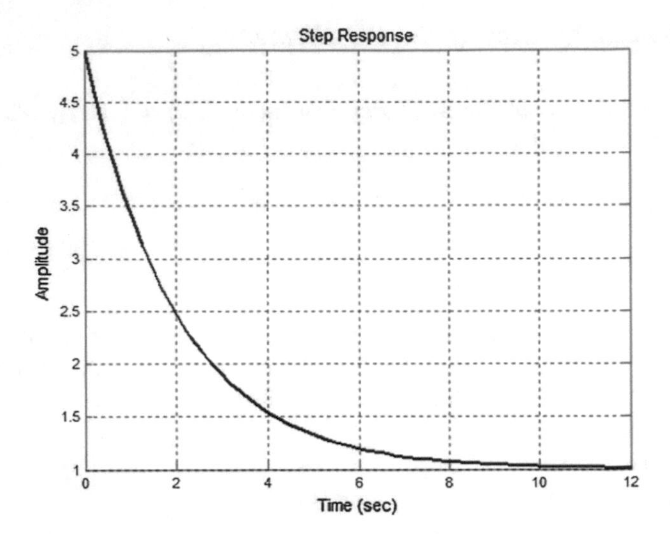

Fig. 8.3 Step response of a *PD* element

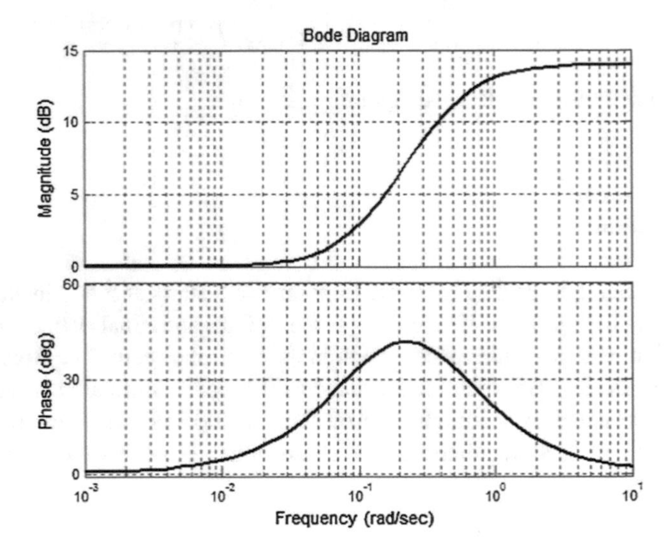

Fig. 8.4 Bode diagram of a *PD* element

the series form is more advantageous: the transfer function of the regulator is approximated by serially connected *PI* and *PD* elements (see formula (8.8) in the textbook [1]).

$$C(s) = A_{\mathrm{P}} \frac{sT_{\mathrm{I}} + 1}{sT_{\mathrm{I}}} \frac{sT_{\mathrm{D}} + 1}{sT + 1}$$

Plot the step response and the Bode diagram of the element.

```
C=(10*s+1)/(10*s)*(s+1)/(0.5*s+1)
t=0:0.1:15;
step(C,t),grid
bode(C),grid
```

In the step response (Fig. 8.5), at the beginning the differentiating effect that is responsible for acceleration is dominant, whereas for later times the integrating effect is dominant. The initial low frequency part of the Bode amplitude diagram (Fig. 8.6) can be approximated by an asymptote of slope −20 dB/decade, and the middle frequency part by an asymptote of slope +20 dB/decade.

8.2 Design of a *PID* Regulator

Consider the closed loop control system shown in Fig. 8.7, with $P(s)$ being the transfer function of the process (plant) to be controlled, and $C(s)$ the transfer function of the regulator.

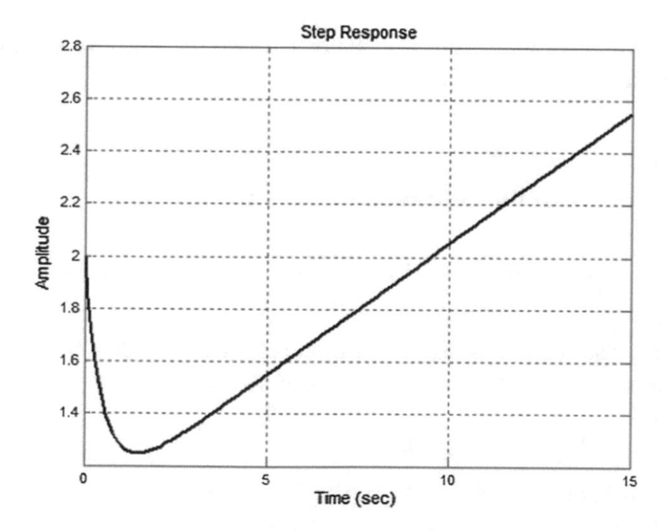

Fig. 8.5 Step response of a *PID* element

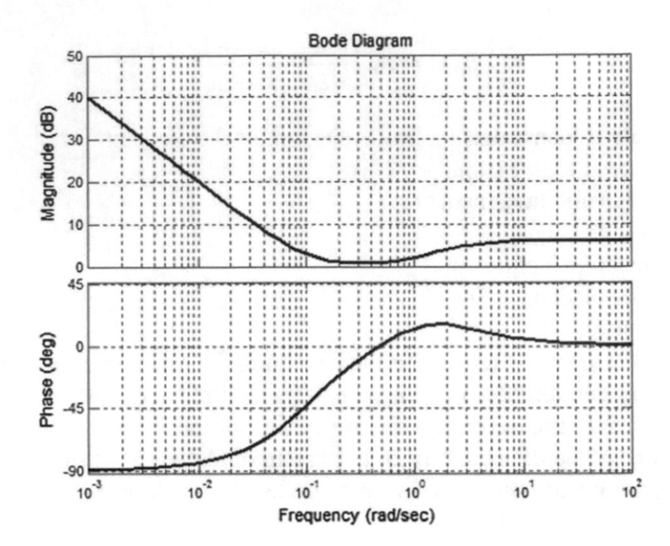

Fig. 8.6 Bode diagram of a *PID* element

Fig. 8.7 Control system

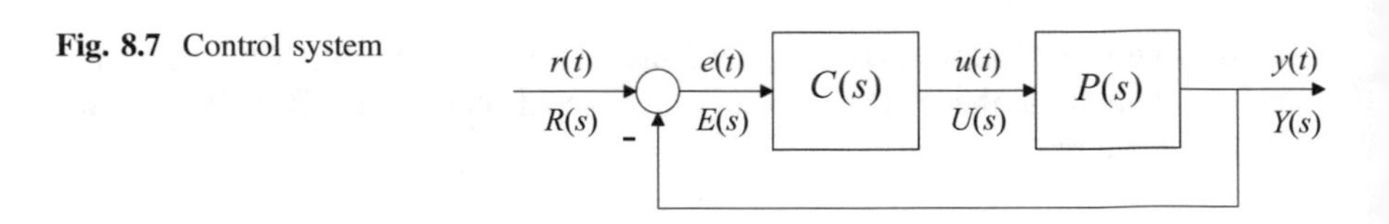

For the given process, a series regulator is to be designed which ensures the fulfilment of the quality specifications.

8.2.1 Design Considerations

The following requirements are set for a closed loop control system:

- Stability;
- Static response (reference signal tracking, disturbance rejection);
- Transient response (overshoot in the output signal, settling time);
- Robustness;
- Limitation of the control signal.

Quality specifications (requirements) can be formulated in the frequency domain as well (see also Chap. 6). The overshoot of the step response of the closed loop is related to the maximum amplification of the amplitude-frequency function of the closed loop, and also to the phase margin calculated from the frequency function of the open loop. If the phase margin is about 60°, the overshoot will be about 10%. The settling time depends on the cut-off frequency, and can be approximated by $\frac{3}{\omega_c} \leq t_s \leq \frac{10}{\omega_c}$.

In practice the manipulated variable, the control signal, is restricted: its value should not exceed a given limit.

The structure and the parameters of the regulator have to be chosen considering the design specifications.

In more complex control problems several restrictions can be imposed on the output and control signals. For example, the integral of the quadratic error and also that of the control signal have to be minimized. The restrictions can be non-linear and also can be contradictory. In general, the parameters of the regulator can be determined by optimization procedures.

In the sequel a simple practical method will be presented for regulator design (also called *compensation*), when the phase margin is prescribed. The design is based on the frequency function of the open loop, and from the properties of the open loop consequences are drawn for the behaviour of the closed loop.

When the requirement is to track the step reference signal accurately, without steady state error, a *PI* regulator is employed. This requirement can be fulfilled by using an integrating effect in the open loop. With a *PD* regulator, the control system can be accelerated. If both the accuracy and the settling time should be improved, a *PID* regulator is employed. The *PD* element causes a significant increase of the initial value of the control signal, which is responsible for the acceleration.

The regulator is designed considering the model of the process and the quality specifications. A usual technique is pole cancellation, when the zeros of the transfer function of the regulator cancel the unfavourable poles of the process, and so a desired dynamics is ensured in the closed loop control circuit. For the pole cancellation technique, it is advantageous to give the transfer function of the regulator in product form.

In a *PID* regulator there are 4 parameters: $k_c = A_P$, T_I, T_D, T. When designing a regulator with the pole cancellation technique, these parameters are chosen as follows: T_I should be equal to the biggest time constant (the pole with the lowest frequency), and T_D should be equal to the second biggest time constant. Thus the zeros introduced by the regulator cancel the poles of the process. The parameter T is given in the form $T = \frac{T_D}{n_p}$, where n_p is the ratio of pole replacement, which indicates how far away the *PD* element pushes the compensated pole of the system. A good experimental rule is to choose the value of n_p in the range 2–10. If it is higher, the control system will be faster, but at the cost of a higher maximum of the control signal. As the value of k_c does not influence the phase of the open loop frequency function, this parameter can be used to set the value of the prescribed phase margin.

The steps of regulator design with *P*, *PI*, *PD* and *PID* regulators are shown for compensating the process given by the transfer function $P(s) = 1/[(1+10s)(1+s)(1+0.2s)]$.

Design series *P*, *PI*, *PD* and *PID* regulators to ensure a phase margin of about 60°. Give the quality characteristics of the compensated system. Calculate and draw the output and the control signals for a unit step reference signal.

8.2.2 Design of a P Regulator

The regulator is given as $C(s) = k_c$. So only the value of k_c has to be determined.

Give the transfer function of the process.

```
s=zpk('s')
P=1/((1+10*s)*(1+s)*(1+0.2*s))
```

First let $k_c = 1$.

```
kc=1
C=kc
L=C*P
```

Draw the Bode diagram of the open loop and determine its characteristic values (phase margin, gain margin, and cut-off frequency) using the command margin.

```
margin(L)
```

The system has significant phase and gain margins. The phase angle decreases monotonically from zero to −270°, so by changing the gain the required phase margin could be set. The gain of the regulator will be the reciprocal of the gain belonging to the phase value $\varphi = \varphi_t - 180°$. This can be read off the Bode diagram of the open loop, or calculated from a table containing the corresponding frequency, gain and phase values, or found by using command margin.

As the calculation of the gain margin g_t is similar to the calculation of k_c (e.g. if $\varphi_t = 60°$).

$$k_c = 1 \Big/ |L(j\omega)|_{\varphi=-120^\circ} \quad \text{and} \quad g_t = 1 \Big/ |L(j\omega)|_{\varphi=-180^\circ},$$

therefor k_c can also be found using the command margin. The input parameter of the command margin can be a transfer function given in *LTI sys* structure, or the gain, phase and frequency vector calculated by the command bode. The command margin calculates the gain margin from the gain belonging to the phase value −180°. If the phase angles are decreased by the value of the required phase margin, then margin will calculate g_t as the reciprocal of the gain belonging to the given phase margin.

```
[mag,phase,w]=bode(L);
gt=margin(mag,phase-60,w)
kc=gt
```

So the regulator is $C_P(s) = k_c = 7.51$. The parameter k_c can also be calculated from the table below containing the corresponding data of the frequency function.

```
Table=[mag(:), phase(:), w]
```

mag	phase	w
0.2756	-95.0730	0.3290
0.1960	-107.0164	0.4520
0.1340	-119.7735	0.6210
0.0873	-133.4679	0.8532

The parameter k_c is calculated as the reciprocal of the gain corresponding to the phase angle $-120°$, and the corresponding frequency is the cut-off frequency: $k_c = 1/0.134 = 7.4627$, $\omega_c = 0.621$.

Refining the resolution of the frequency vector w, the two methods give the same result.

Similarly to the method using the table above, k_c can be found also from the Bode diagram.

```
bode(L)
```

Change the scale of the amplitude curve from decibels to *absolute*: clicking on the white background of the Bode amplitude diagram, choose with the right mouse button *Properties –> Units –> Magnitude in—absolute*. Find phase angle $-120°$ in the phase diagram then read off the gain belonging to this frequency from the Bode amplitude diagram. The reciprocal of this gain will be k_c, i.e. $k_c = \frac{1}{|L(j\omega)|_{\varphi=-120°}} = \frac{1}{0.134} = 7.46$.

Check the behaviour of the system.

```
C=kc
L=kc*L
```

Check the parameters characterizing the stability margins (the gain margin and phase margin).

```
margin(L)
[gt,pm,wg,wc]=margin(L)
```

The phase margin is indeed 60°. The cut-off frequency is $\omega_c = 0.6245$.

Calculate the resulting transfer function of the closed loop system.

```
T=L/(L+1)
```

The calculations may result in coinciding zero-pole pairs, which can be cancelled using the command minreal.

```
T=minreal(T)
```

Or, in one step,

```
T=minreal(L/(1+L))
```

The same result is obtained using the command feedback.

```
T=feedback(L,1)
```

Plot the Bode diagrams of the open and the closed loop on one diagram.

```
bode(L,'r',T,'b')
```

It can be seen that in the low frequency range the amplitude diagram of the closed loop is approximately 1, and in the high frequency range the two curves are approximately the same.

Plot the step response of the closed loop.

```
step(T)
```

Calculate its values.

```
t=0:0.05:10;
y=step(T,t);
```

The maximum value of the step response:

```
ym=max(y)
```

The steady state value of the step response:

```
ys=dcgain(T)
```

From these values the overshoot can be calculated as

```
yt=(ym-ys)/ys
```

The error in steady state:

```
es=1-ys
```

Let us analyse the behaviour of the control signal $u(t)$. This is important because this is the input of the process, and it is not allowed to exceed the given limits. Let us calculate the resulting transfer function between the control signal and the reference signal.

```
U=minreal(C/(1+L))
```

or

```
U=feedback(C,P)
```

For a step reference signal:

```
ut=step(U,t);
plot(t,ut)
```

The constraint is generally imposed as the maximum value of the control signal.

```
um=max(ut)
```

8.2.3 Design of P, PI, PD and PID Regulators

PI, *PD* and *PID* regulators can be designed similarly. Let us apply the pole cancellation technique. The table below summarizes the structure of the regulators and the characteristic values of the control systems with the different regulators. Let us generate a new *m*-file for regulator design. The program calculates the gain k_c of that regulator which ensures a phase margin of 60°, then evaluates the characteristic parameters of the control system. In the table yt denotes the overshoot of the output signal, es is the value of the static error for unit step reference signal, um denotes the maximum value of the control signal, and ts gives the settling time.

To write a MATLAB™ program, let us generate a new *m*-file. This text file can be opened in the file menu of MATLAB™ with the extension ".*m*". Write the MATLAB™ commands into the empty file. Then save it: *Save As, C:/Matlab/work/myfile.m.*

To call the program simply write the name of the program, without its extension, in the command window of MATLAB™.

```
myfile
```

The following MATLAB™ program realizes the regulator design.

```
clear; s=zpk('s');
P=1/((1+10*s)*(1+s)*(1+0.2*s))
Cp=1; Cpi=(1+10*s)/(10*s)
Cpd=(1+s)/(1+0.2*s)
Cpid=Cpi*Cpd
[mag,phase,w]=bode(Cp*P);
kp=margin(mag,phase-60,w)
Cp=kp*Cp;

[mag,phase,w]=bode(Cpi*P);
kpi=margin(mag,phase-60,w);
Cpi=kpi*Cpi;

[mag,phase,w]=bode(Cpd*P);
kpd=margin(mag,phase-60,w);
Cpd=kpd*Cpd;

[mag,phase,w]=bode(Cpid*P);
kpid=margin(mag,phase-60,w);
Cpid=kpid*Cpid;

Lp=Cp*P;
Lpi=minreal(Cpi*P,0.0001)
Lpd=minreal(Cpd*P,0.0001)
Lpid=minreal(Cpid*P,0.0001)
%Resulting transfer functions:
Tp=Lp/(1+Lp);
% or Tp=feedback(Lp,1);
Tpi=Lpi/(1+Lpi);
Tpd=Lpd/(1+Lpd);
Tpid=Lpid/(1+Lpid);
%Transfer functions of U(s):
Up=Cp/(1+Lp);
Upi=Cpi/(1+Lpi);
Upd=Cpd/(1+Lpi);
Upid=Cpid/(1+Lpid);
t=0:0.05:10;

figure(1),step(Tp,'r',Tpi,'b'
,Tpd,'g',Tpid,'m',t)
figure(2),step(Up,'r',Upi,'b'
,t)
figure(3),step(Upd,'g',Upid,'m'
,t)

yp=step(Tp,t);
ypi=step(Tpi,t);
ypd=step(Tpd,t);
ypid=step(Tpid,t);

ysp=dcgain(Tp)
yspi=dcgain(Tpi)
yspd=dcgain(Tpd)
yspid=dcgain(Tpid)
ep=1-ysp
epi=1-yspi
epd=1-yspd
epid=1-yspid

ytp=(max(yp)-ysp)/ysp
ytpi=(max(ypi)-yspi)/yspi
ytpd=(max(ypd)-yspd)/yspd
ytpid=(max(ypid)-yspid)/yspid
up=step(Up,t);
upi=step(Upi,t);
upd=step(Upd,t);
upid=step(Upid,t);
upim=max(upi)
updm=max(upd)
upidm=max(upid)
```

The step responses are shown in Fig. 8.8. The control signals for the *P* and *PI* controls are given in Fig. 8.9. The control signals for the *PD* and *PID* controls are shown in Fig. 8.10.

The following table summarizes the structure and the parameters of different regulators given in the MATLAB™ example, and presents the characteristic values of the closed system.

	$C(s)$	$L(s)=C(s)\,P(s)$	k_c	ω_c	yt	es	um	~ts
	No control case	$\frac{1}{(1+10s)\,(1+s)\,(1+0.2s)}$		0	0	0.5	2	12
P	k_c	$\frac{k_c}{(1+10s)\,(1+s)\,(1+0.2s)}$	7.51	0.62	0.153	0.117	7.5	8
PI	$k_c\frac{1+10s}{10s}$	$\frac{k_c}{s(1+s)\,(1+0.2s)}$	5.04	0.46	0.078	0	5.16	9
PD	$k_c\frac{1+s}{1+0.2s}$	$\frac{k_c}{(1+10s)(1+0.2s)^2}$	16.5	1.51	0.103	0.057	82.7	2
PID	$k_c\frac{(1+10s)(1+s)}{10s\,(1+0.2s)}$	$\frac{k_c}{s(1+0.2s)^2}$	14.3	1.33	0.076	0	71.3	2

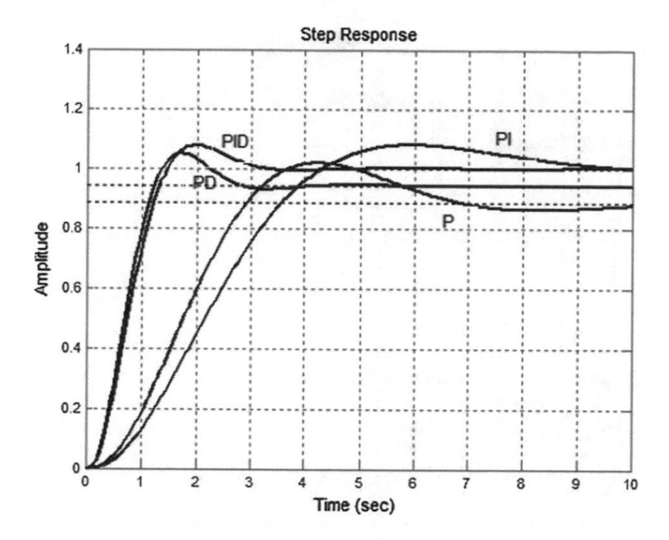

Fig. 8.8 Step responses of the control system in case of *P*, *PI*, *PD* and *PID* controllers

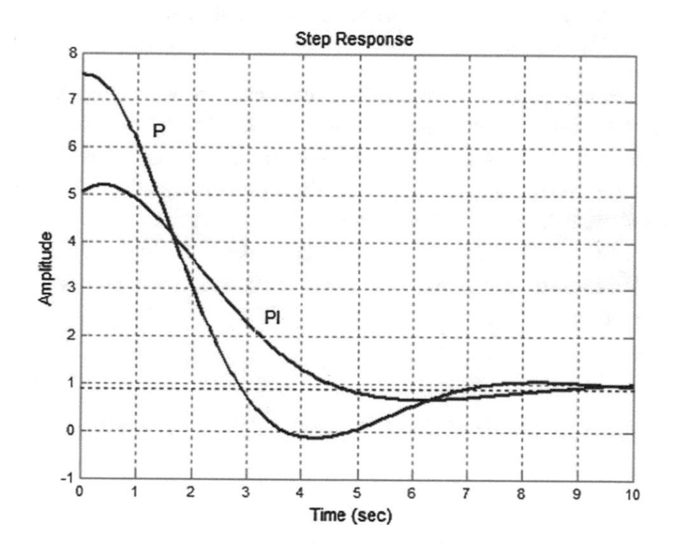

Fig. 8.9 The control signals in case of *P* and *PI* controllers

The requirements imposed on the control system are a fast settling process and good reference signal tracking. In Fig. 8.8 it can be seen that *P* compensation does not fulfill these conditions. The settling is slow and the output signal does not reach

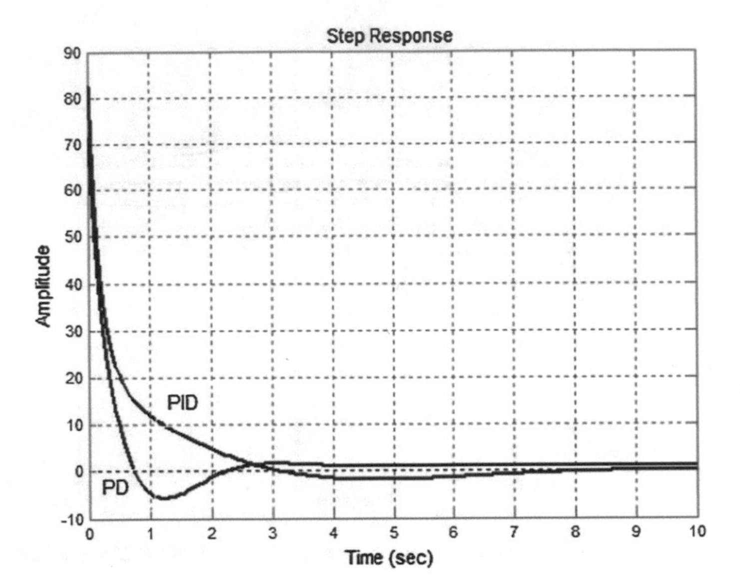

Fig. 8.10 The control signals in case of *PD* and *PID* controllers

the required value $y(\infty) = 1$. With *PI* compensation the static error has been decreased to zero, but the settling process is slow. The *PD* compensation accelerates the control system, but there is a static error. The reason for this acceleration (Fig. 8.10) is the significant increase of the control signal $u(t)$. With the *PID* regulator the system became fast and the static error is zero.

The behaviour of the control system can also be analysed by building a SIMULINK™ block diagram and running it with the given process and with the designed regulators.

8.2.4 Regulator Design for a Second-Order Oscillating Element

The process is given by the following transfer function:

$$P(s) = \frac{A'}{(s-p_1)(s-p_2)} = \frac{A}{s^2T_o^2 + 2\xi T_o s + 1}, \text{ where } p_{1,2} = a \pm jb$$

The poles are complex conjugates. The breakpoint frequency of the Bode diagram is $\omega_o = 1/T_o$. Here the slope of the approximate Bode amplitude diagram changes from 0 to −40 dB/decade. In this case a possible *PID* pole cancellation technique can be, if the time constants of both the *PI* and the *PD* elements are chosen to be the reciprocals of the natural frequency, i.e. $T_I = T_o$ and $T_D = T_o$, so

$$C(s) = k_c \frac{1 + T_o s}{s} \frac{1 + T_o s}{1 + T_1 s}.$$

Another possibility is to employ a pure integrating element as a regulator, whose gain is set to ensure a phase margin of about 60°. Let us remark that for a small damping factor ξ, the prescribed phase margin alone will not always ensure the appropriate transient response. It is necessary to arrange that in the vicinity of the cut-off frequency the BODE amplitude diagram does not move close to the 0 dB axis.

8.2.5 *Applying Experimental Tuning Rules*

Besides the discussed regulator design methods, there are several practical regulator tuning methods. The most frequently used methods which give rules of thumb for the parameter tuning of the *PID* regulator based on the model of the process are:

- The ZIEGLER-NICHOLS rules
- The OPPELT method
- The CHIEN-HRONES-RESWICK method
- The STREJC method
- The ÅSTRÖM relay method
- The ÅSTRÖM-HÄGGLUND method.

The rules of thumb are given in the textbook [1].

8.3 *PID* Regulator Design for a Dead-Time System

Compensating a system containing dead-time is more complicated than for a system without dead-time, as the transfer function of the dead-time element can not be represented accurately by a rational function. The phase shift caused by the dead-time has to be taken into consideration in regulator design.

Let us consider the control system given in Fig. 8.11.

Here $P(s) = P_+(s)e^{-sT_d}$, where $P_+(s)$ is the transfer function of the process without the dead-time and T_d denotes the dead-time, $C(s)$ is the transfer function of the regulator, and $L(s) = C(s)P(s)$ is the loop transfer function.

Fig. 8.11 Control system

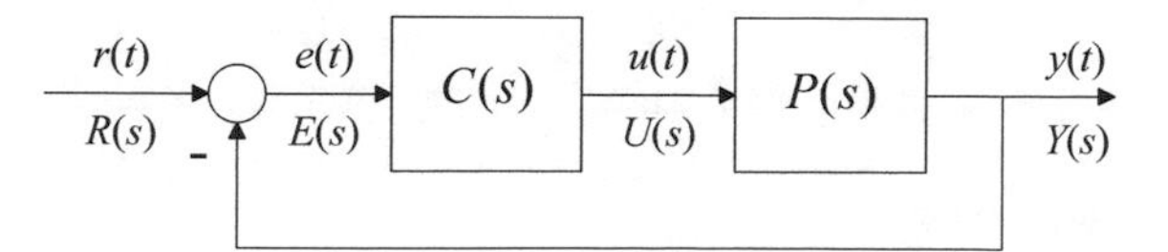

Consider the following example:

$$P(s) = P_+(s)\,\mathrm{e}^{-sT_\mathrm{d}} = \frac{\mathrm{e}^{-s}}{1+20s}.$$

The transfer function of the regulator $C(s)$ has to be chosen to ensure the fulfilment of the quality specifications.

Prescriptions: For a step reference signal, the static error for reference signal tracking should be zero, and the overshoot of the output signal should be below 10%. These requirements can be ensured using a *PI* regulator: $C(s) = k_\mathrm{c}(1+20s)/s$.

The loop transfer function is

$$L(s) = C(s)P(s) = k_\mathrm{c}\frac{1+20s}{s}\frac{\mathrm{e}^{-s}}{1+20s} = k_\mathrm{c}\frac{\mathrm{e}^{-s}}{s}.$$

The constant k_c is chosen to ensure a phase margin of about 60°.

The amplitude of the frequency function of the process is calculated from its part without the dead-time:

$|P(j\omega)| = |P_+(j\omega)\mathrm{e}^{-j\omega T_\mathrm{d}}| = |P_+(j\omega)|$, as $|e^{-j\omega T_\mathrm{d}}| = 1$

The phase angle is

$$\arg\{P(j\omega)\} = \arg\{P_+(j\omega)\} + \arg\{\mathrm{e}^{-j\omega T_\mathrm{d}}\} = \arg\{P_+(j\omega)\} - \omega T_\mathrm{d}$$

Regulator design in the MATLAB™ environment can be executed in two ways. In the first way, in the frequency domain the phase angle of the process without the dead-time is calculated, and then it is modified by $-\omega T_\mathrm{d}$, the phase angle of the dead-time element. The disadvantage of this method is that the simulation can not be done in the MATLAB™ environment alone, analysis in SIMULINK™ is also required. In the second way the dead-time is approximated by a rational function. For the approximate process the regulator design can be executed according to the method applied for rational functions (see Sect. 8.2 in textbook [1]). With this method, the behaviour of the system can be analysed in the MATLAB™ environment.

Let us emphasize that the first method is preferred.

8.3.1 *Regulator Design for a Dead-Time System Considering the Phase Shift*

Let us design a *PI* regulator for the process given in Sect. 8.3. Write $P_+(s)$ = P.

```
s=zpk('s')
P=1/(1+20*s)
Td=1
```

```
kc=1
C=kc*(1+20*s)/s
```

The transfer function of the open loop is

```
L=C*P,L=minreal(L)
```

The amplitude-frequency function of the open loop coincides with the amplitude of the system without dead-time. Its phase angle is modified by a linear term. To execute the calculations a frequency vector w has to be defined, which contains the cut-off frequency. In many cases the command bode itself calculates the frequency vector.

```
[mag,phase,w]=bode(L);
```

If this frequency vector is not satisfactory, the user has to define it so that it ensures a frequency range which is wide enough. This can be done using the knowledge of the process or by trial and error.

```
w=logspace(-1,1,100)'
[mag,phase]=bode(L,w);
magd=mag(:);
phased=phase(:)-w*Td*180/pi;
```

The gain k_c can be calculated in two ways.

1. *Method 1*: with the command margin.

The gain margin for a phase angle $-120°$ is found by

```
gm=margin(magd,phased-60,w)
```

This will be the value of gain a k_c:

```
kc=gm
      0.5235
```

2. *Method 2*: using a table.

```
Tabl=[phased, magd, w]
   -116.5943   2.1544   0.4642
   -117.8607   2.0565   0.4863
   -119.1873   1.9630   0.5094
   >> -120.5770   1.8738   0. 5337  <<
   -122.0330   1.7886   0.5591
   -123.5583   1.7074   0.5857
```

The value of magd at phased $= -120$ is 1.8738.
Hence $k_c = 1/1.8738$ and $\omega_c = 0.5337$.

```
kc=1/1.8738
        0.5337
```

Let us calculate the regulator transfer function again:

```
C=kc*(1+20*s)/s
L=C*P,L=minreal(L)
```

Check the phase margin.

```
[mag,phase]=bode(L,w);
magd=mag(:);
phased=phase(:)-w*Td*180/pi;
margin(magd,phased,w)
```

Figure 8.12 shows the Bode diagram of the open loop and the phase margin. In this simple case, the calculation can also be executed analytically. The transfer function of the open loop is $1/s$, its phase angle is $-90°$ over the whole frequency range. At the prescribed phase margin of 60°, the phase angle is $-120°$. That means that the dead-time can add $-30°$, i.e. $-\omega_c T_d = -\pi/6$. Hence $\omega_c = \pi/6 = 0.5236$. From the condition $1 = |k_c e^{-j\omega}/j\omega|$ the gain is calculated as $k_c = \omega_c = 0.5236$, which is close to the result obtained previously from the table.

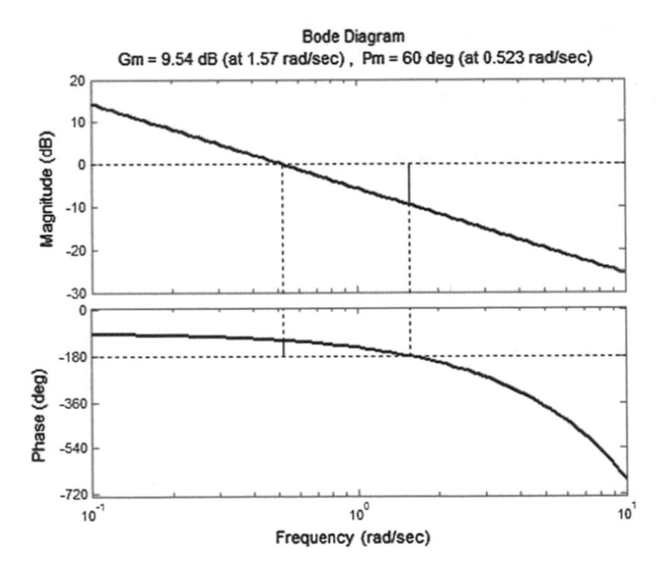

Fig. 8.12 Bode diagram of the open loop

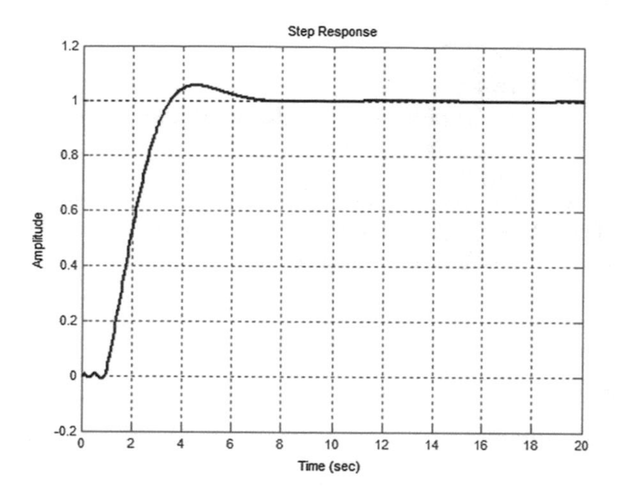

Fig. 8.13 Step response of the control system

The behaviour of the system can be analysed with the SIMULINK™ model shown in Fig. 8.14. With SIMULINK™, the dead-time can be simulated easily with the *Transport Delay* block. Set its parameter to the value Td, and the transfer function of the process to P. Figure 8.13 shows the step response of the control system.

The result of the simulation can be sent to the MATLAB™ surface for further analysis and graphical representation. This can be done in two ways: with the *To Workspace* block, or with the *Scope* block. In the *To Workspace* block the name of the variable that will be used in MATLAB™ has to be set, its type has to be given as *Matrix*. Then the signals can be plotted using MATLAB™:

```
plot(t,y),grid
```

In MATLAB™ the results of the simulation can also be analysed and plotted from *Scope* blocks. Set the parameters of the graphical window of *Scope* as follows:

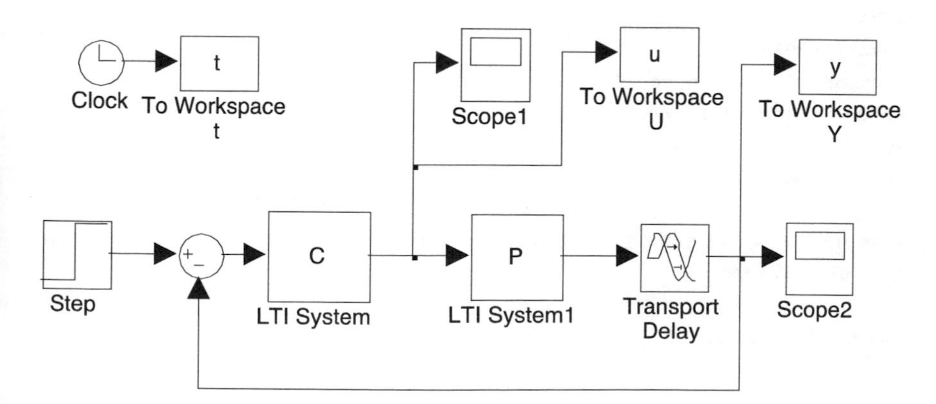

Fig. 8.14 SIMULINK™ model of a control system with dead time in the process

Under the *properties* menu,

Data history: *Save data to workspace*
Variable name: ty (tu in the case of the control signal)
Matrix format

So the vectors of the time t and the output y can be obtained easily after the simulation. Then the quality characteristics (overshoot, settling time, maximum value of the control signal) can be determined.

```
t=ty(:,1)
y=ty(:,2)
plot(t,y),grid
```

8.3.2 *Regulator Design for a Dead-Time System Using Pade Approximation*

The transfer function of a dead-time element can be approximated by the Pade rational function, $P_{\mathrm{Pade}}(s) \cong e^{-sT_{\mathrm{d}}}$. (The first few terms of the Taylor series of the Pade rational function are the same as those of the transfer function of the dead-time element.)

$$P(\mathrm{s}) = P_{+}(\mathrm{s})\, e^{-sT_{\mathrm{d}}} \cong P_{+}(\mathrm{s})\, P_{\mathrm{Pade}}(s)$$

In MATLAB™, the command pade calculates the approximation for the given degree. For example, for the 5th degree:

```
s=zpk('s')
P=1/(1+20*s)
Td=1
kc=1
C=kc*(1+20*s)/s
[numpade,denpade]=pade(Td,5)
Ppade=tf(numpade,denpade)
Ppade=zpk(Ppade)
Pd=P*Ppade
```

From this point on, the steps of the regulator design are the same as for a system without dead-time.

```
L=C*Pd,L=minreal(L)
```

The gain k_c is calculated by

```
[mag,phase,w]=bode(L);
kc= margin(mag,phase-60,w)
          0.5212
C=kc*(1+20*s)/s; L=kc*L;
T=L/(1+L); T=minreal(T)
```

The settling time can be evaluated with the command step. The step response can be compared with the result obtained with SIMULINK™ (Fig. 8.15).

```
step(T,20),grid
```

It can be seen that the step responses obtained by the two methods are approximately the same.

It has to be emphasized that the method given in Sect. 8.3.1 is more accurate than the method using the Padé approximation. With the Padé approximation, the transfer functions of the open and the closed loop are complicated because of the high degree approximation. The advantage of Padé approximation is that the design method is the same as for systems without dead-time.

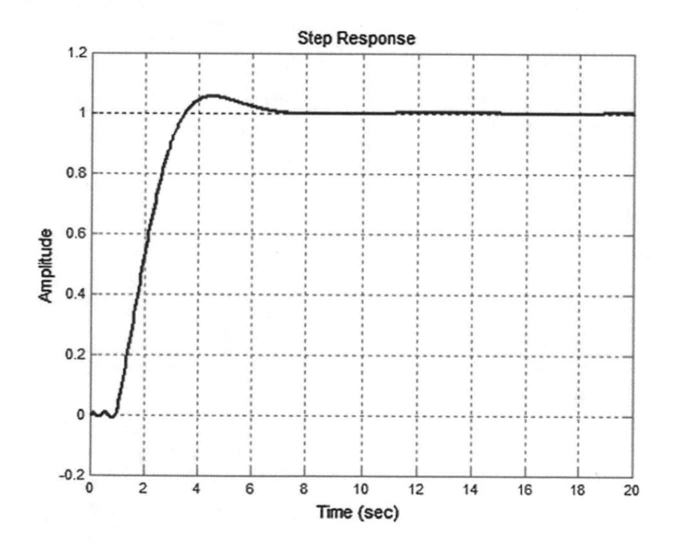

Fig. 8.15 Step response calculated with Padé approximation of the dead time

8.4 Control of an Unstable System

8.4.1 Control of an Unstable System with a P Regulator

An unstable process is given by the transfer function $P(s) = 20/[(s+2)(s-5)]$. Let us analyse whether the process can be stabilized with a proportional regulator $C(s) = k_c$ or not.

```
s=zpk('s')
P=20/((s+2)*(s-5))
figure(1); grid on; nyquist(P);
figure(2); grid on; bode(P);
figure(3);rlocus(P);
```

The transfer function of the process has a pole in the right half-plane. According to the general Nyquist stability criterion the control system can be stabilized with the given regulator, if the Nyquist diagram of the open loop encircles $-1+j0$ counter-clockwise as many times as the number of the poles of the process in the right half-plane. Figure 8.16 shows that the control system can not be stabilized, as the Nyquist diagram encircles $-1+j0$ clockwise. The Bode diagram shows that the phase margin is negative. The same result is obtained by calculating the roots of the characteristic equation $s^2 - 3s - 10 + k_c = 0$. The necessary condition of stability is that the coefficients are of the same sign, which is not fulfilled. The root locus $(0 < k_c < \infty)$ gives the same result. At each value of the gain, at least one root of the characteristic equation lies in the right half-plane.

Let us consider the process given by $P(s) = 5/[(s-2)(s+5)]$. Can this process be stabilized with a proportional $C(s) = k_c$ regulator?

```
clear
s=zpk('s')
```

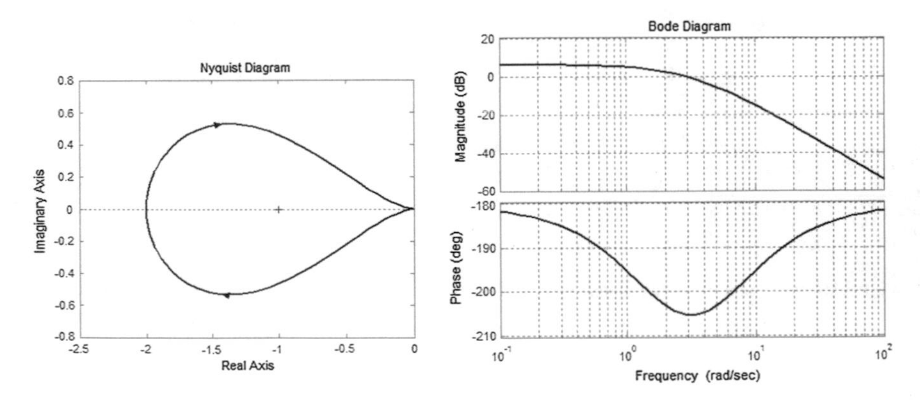

Fig. 8.16 Nyquist and Bode diagrams of an unstable process controlled by *P* regulator

```
P=5/((s-2)*(s+5))
figure(1);grid on;nyquist(P);
figure(2); grid on; bode(P);
figure(3); rlocus(P);
```

As can be seen from Fig. 8.17, the NYQUIST diagram may encircle the point $-1+j0$ counter-clockwise. This can be arranged by increasing the gain. For $k_c > 2$ the curve encircles the point $-1+j0$, so the control system becomes stable. The BODE diagram shows that the phase margin can be positive.

Let us choose the gain k_c to ensure that the cut-off frequency is located where the phase margin is of maximum value. First analyse the behaviour of the open loop for $k_c = 1$.

```
C=1
L=C*P
[mag,phase,w]=bode(L);
```

Method 1: using the table

```
T=[phase(:), mag(:), w]
   -158.1444    0.3559    1.7433
   -155.9825    0.3066    2.2122
   -154.7797    0.2530    2.8072
>> -154.6231    0.2259    3.1623   <<
   -154.7797    0.1994    3.5622
   -155.9825    0.1501    4.5204
```

The BODE phase curve reaches its maximum at `phase` = 154.62, `mag` = 0.2259, $\omega = 3.16$. With a proportional regulator, the maximum reachable phase margin is

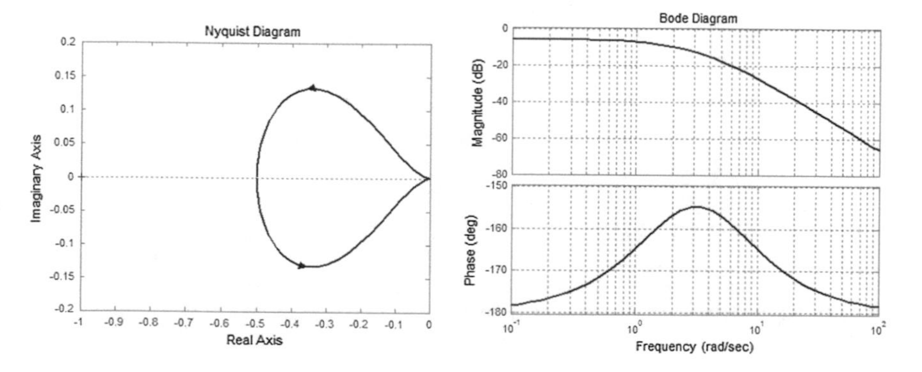

Fig. 8.17 Nyquist and Bode diagrams of an unstable process controlled by *P* regulator

$\varphi_t = 180° - 154.6° = 25.8°$. In this case the gain has to be chosen as $k_c = 1/0.045 = 4.42$.

```
kc=4.42
```

Method 2: The maximum value can be calculated with the command max.

```
[maxphase,index]=max(phase)
kc=1/mag(index)
```

Calculate the regulator again:

```
C=kc;
L=C*P;
```

With margin, the phase margin can be checked graphically.

```
margin(L);
```

The phase margin of the system is small, $\varphi_t = 25.4°$ (60° would be required). Plot the step response of the closed loop (Fig. 8.18.).

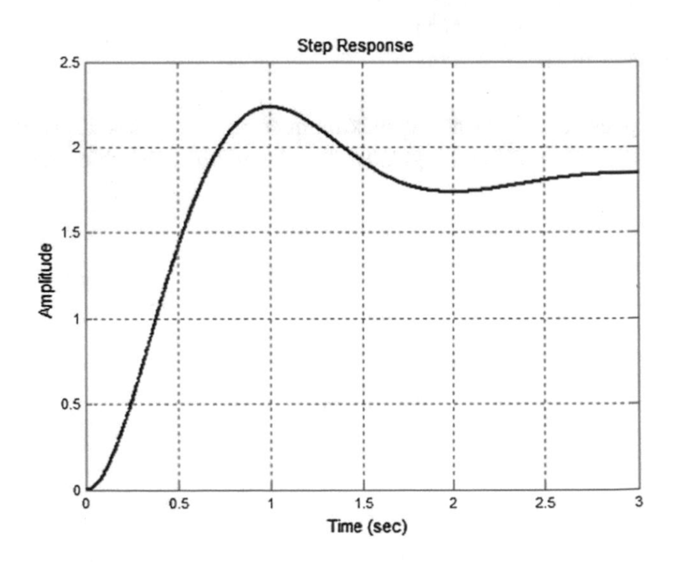

Fig. 8.18 Step response of *P* control of an unstable process with maximum phase margin

```
T=feedback(L,1)
step(T), grid
```

Stable behaviour has been reached, but there is a quite big overshoot and there is a significant static error. Applying a *PID* regulator the performance of the control system could be improved.

8.4.2 Control of an Unstable System with a PID Regulator

To decrease the static error, let us design a *PID* regulator.

$$C(s) = k_c \frac{s+2}{s} \frac{s+5}{s+50}.$$

The unstable pole $p_1 = 2$ can not be cancelled by a zero, as the parameters generally are obtained from measured data, and the system would become unstable even in the case of a small difference between the pole and the cancelling zero. On the other hand with pole cancellation of the unstable pole the inner stability can not be ensured, as the unstable pole would appear in the resulting transfer function between the output and the inner disturbance signals. The unstable pole could be compensated by a stable *PI* element. To accelerate the system the stable pole $p_1 = -5$ is shifted to a higher frequency by a *PD* element ($p = -50$, the pole shift ratio is 10). The gain k_c is chosen again to ensure the maximum phase margin.

Clear all the variables and close the graphic windows.

```
clear all, close all
s=zpk('s')
P=5/((s-2)*(s+5))
C=((s+2)*(s+5))/(s*(s+50))
L=C*P
L=minreal(L)
bode(L)
[mag,phase,w]=bode(L);
```

Determine the gain for the maximum phase:

```
[maxphase,index]=max(phase)
kc=1/mag(index)
```

The gain is $k_c = 152$, and the phase margin $\varphi_t = 180 + \mathsf{maxphase} = 58°$. Check the behaviour of the control system.

```
C=kc*((s+2)*(s+5))/(s*(s+50))
L=C*P, L=minreal(L)
margin(L)
```

The obtained phase margin is really 60° (Fig. 8.19).
Determine the step response of the closed loop and the control signal (Fig. 8.20).

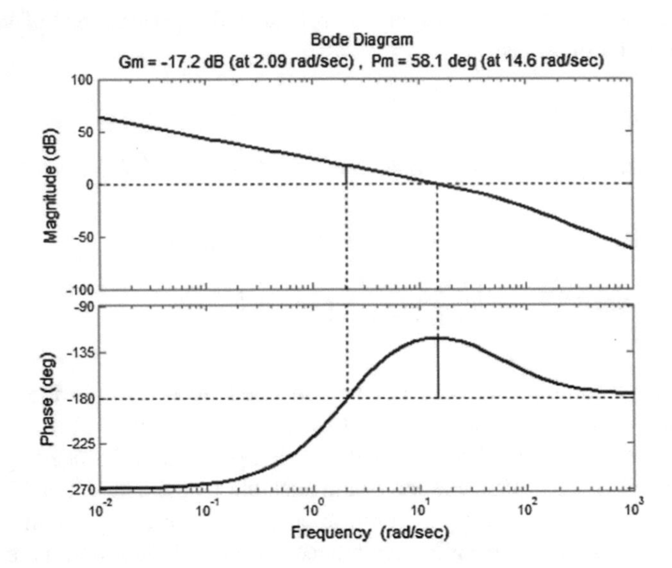

Fig. 8.19 BODE diagram of *PID* control of an unstable process with phase margin of 60°

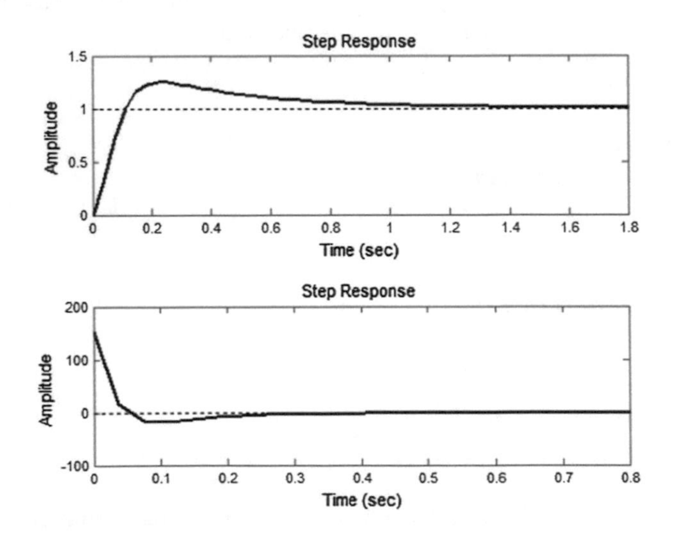

Fig. 8.20 The output and the control signals

```
T= L/(1+L), T=minreal(T)
```

The static error has been decreased to zero.

```
es=1-dcgain(T)
   es = 1.2212e-015
```

The resulting transfer function between the control signal and the reference signal is

```
U=C/(1+L), U=minreal(U)
subplot(211),step(T)
subplot(212),step(U)
u=step(U)
um=max(u)
```

The maximum value `um` of the control signal is high. This value can be decreased by decreasing the pole shift ratio.

8.5 Handling of Constraints

Let us analyse the behaviour of the *PI* regulator designed in Sect. 8.2.3 when there are constraints.

$$P(s) = \frac{1}{(1+10s)(1+s)(1+0.2s)};\ C(s) = 5.04\frac{1+10s}{10\,s}$$

The maximum value of the control signal is $\mathtt{um} = 5.04$. In practical applications limitations do exist for the control signal. Such limitations may originate from several sources. The manipulator which provides the control signal to the process input generally can not produce a higher value than its given maximum. Limitation is applied also at the process input when the process should be protected against too big, harmful interventions.

In the case of *PI* regulators, the FOXBORO regulator provides a simple solution for handling limitations. The regulator is realized by a saturation block fed back with positive feedback by a first order lag element (see Fig. 8.22). Without the saturation the proportional path has gain 1. The resulting transfer function of this circuit provides a *PI* regulator (Fig. 8.21).

If the regulator works in the linear range, then this relationship holds, otherwise its output is limited. Compare the simple limitation at the process input realized by cutting the input signal with the effect of the FOXBORO regulator. The comparison of the regulators is executed in the SIMULINK™ environment.

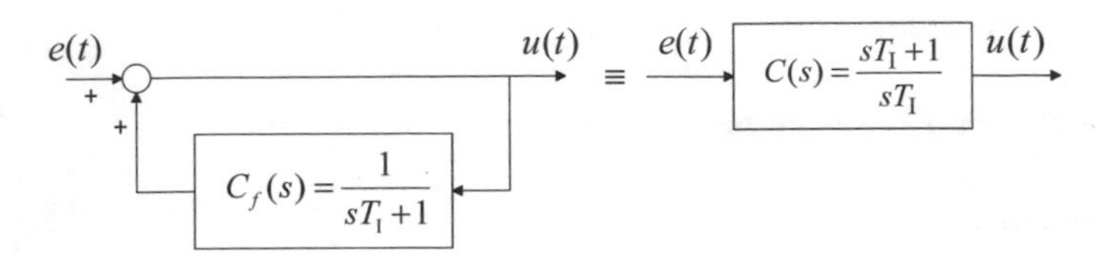

Fig. 8.21 FOXBORO regulator

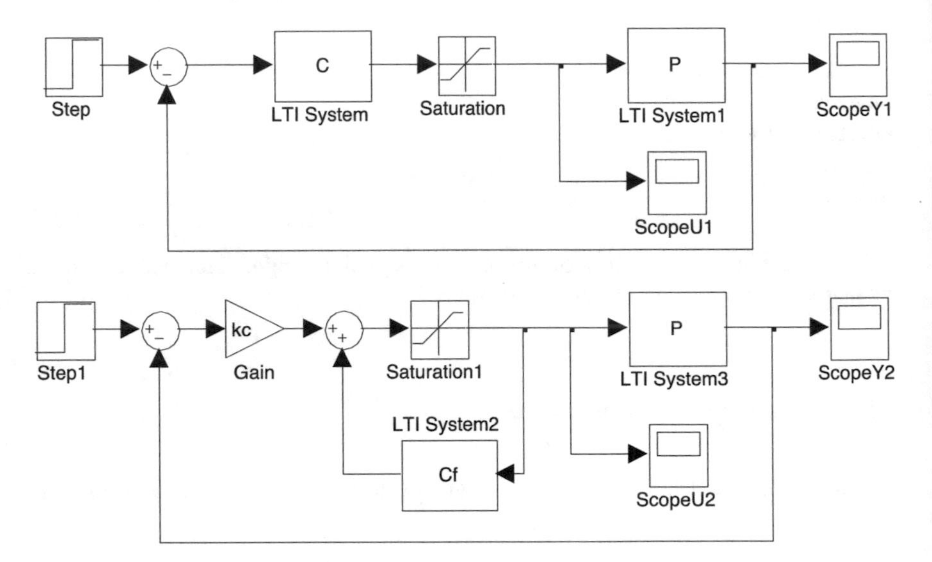

Fig. 8.22 SIMULINK™ diagram of a control system handling saturation by FOXBORO regulator

```
s=zpk('s')
P=1/((1+10*s)*(1+s)*(1+0.2*s))
kc=5.04
C=kc*(1+10*s)/(10*s)
Cf=1/(1+10*s)
```

Build the SIMULINK™ block diagram shown in Fig. 8.22.

Set the lower and upper limits of blocks *saturation* (SIMULINK™ –> *Discontinuities –> Saturation*) (*Upper limit and Lower limit*) to u1 and –u1. Set the simulation time to 50.

```
u1=2
```

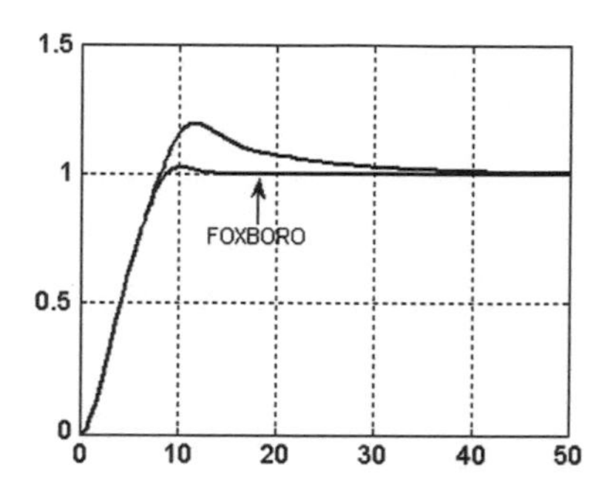

Fig. 8.23 Step responses in case of saturation

In the case of saturation, the course of the output signal with the FOXBORO regulator is more advantageous, the overshoot is smaller, and the settling time is also smaller (Fig. 8.23).

The reason is that with the conventional *PI* control shown in the upper part of Fig. 8.22 at $t = 0$, a control signal of value 5.04 does appear, the saturating element limits this value, and at the output of the regulator the value of the signal will be 2. The saturation quasi “opens” the circuit until the feedback signal brings the saturating element outside of the range of saturation. The output of the integrating element of the *PI* regulator “winds up”, and therefore the saturating element remains in the saturation range for a longer time. In the FOXBORO regulator the saturation acts on the process and on the dynamic part of the regulator located in its feedback path in the same way, therefore the disadvantageous windup phenomenon does not show up here.

Chapter 9
State Feedback Control

Consider first continuous systems. The state space representation of a continuous, linear, time invariant single input–single output system can be given by parameter matrices $\boldsymbol{A}$, $\boldsymbol{b}$, $\boldsymbol{c}$, d in the following form:

$$\dot{\boldsymbol{x}} = \boldsymbol{A}\boldsymbol{x} + \boldsymbol{b}u$$
$$y = \boldsymbol{c}^{\mathrm{T}}\boldsymbol{x} + du$$

(The upper index T indicates transpose, i.e. $\boldsymbol{c}^{\mathrm{T}}$ is a row vector.) The equations above (the state equation and the output equation) determine the transfer function between the u input signal and the y output signal, which is calculated by

$$P(s) = \frac{Y(s)}{U(s)} = \boldsymbol{c}^{\mathrm{T}}(s\boldsymbol{I} - \boldsymbol{A})^{-1}\boldsymbol{b} + d$$

The system model characterized by the four parameters $\{\boldsymbol{A}, \boldsymbol{b}, \boldsymbol{c}, d\}$ is called the state model.

The poles of the model are the roots of the characteristic equation

$$\det(s\boldsymbol{I} - \boldsymbol{A}) = 0.$$

In most practical cases, $d = 0$.

By state feedback, the control signal is obtained from the state variables feeding them back to the input through the constant elements of the vector $\boldsymbol{k}^{\mathrm{T}}$:

$$u = k_{\mathrm{r}}\boldsymbol{r} - \boldsymbol{k}^{\mathrm{T}}\boldsymbol{x}.$$

The state feedback control shown in Fig. 9.1 modifies both the static and the dynamic response of the system between the reference signal r and the output signal y.

L. Keviczky et al., *Control Engineering: MATLAB Exercises*,
Advanced Textbooks in Control and Signal Processing,
https://doi.org/10.1007/978-981-10-8321-1_9

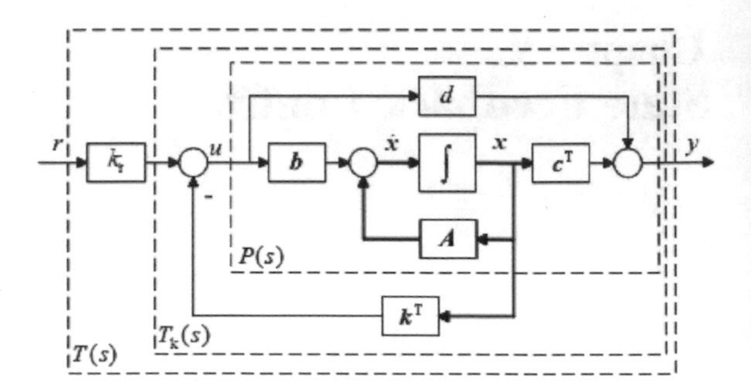

Fig. 9.1 State feedback control

In the feedback, let us consider the feedback (row) vector $\boldsymbol{k}^{\mathrm{T}}$, and in the forward path suppose a compensation factor k_r. The control signal is obtained as $u = k_r\boldsymbol{r} - \boldsymbol{k}^{\mathrm{T}}\boldsymbol{x}$. The equations of the closed loop control system are as follows (here d is also considered, its value is generally zero):

$$\dot{\boldsymbol{x}} = \left(\boldsymbol{A} - \boldsymbol{b}\boldsymbol{k}^{\mathrm{T}}\right)\boldsymbol{x} + k_r\boldsymbol{b}r$$
$$y = \left(\boldsymbol{c}^{\mathrm{T}} - d\boldsymbol{k}^{\mathrm{T}}\right)\boldsymbol{x} + dk_r r$$

By introducing the notation $\boldsymbol{A}_{\mathrm{k}} = \boldsymbol{A} - \boldsymbol{b}\boldsymbol{k}^{\mathrm{T}}$, $\boldsymbol{b}_{\mathrm{k}} = k_r\boldsymbol{b}$, $\boldsymbol{c}_{\mathrm{k}} = \boldsymbol{c} - d\boldsymbol{k}$, $d_{\mathrm{k}} = dk_r$, we have

$$\dot{\boldsymbol{x}} = \boldsymbol{A}_{\mathrm{k}}\boldsymbol{x} + \boldsymbol{b}_{\mathrm{k}}r$$
$$y = \boldsymbol{c}_{\mathrm{k}}^{\mathrm{T}}\boldsymbol{x} + d_{\mathrm{k}}r$$

and the characteristic equation is

$$\det(s\boldsymbol{I} - \boldsymbol{A}_{\mathrm{k}}) = \det\left(s\boldsymbol{I} - \boldsymbol{A} + \boldsymbol{b}\boldsymbol{k}^{\mathrm{T}}\right) = 0.$$

Comparing the characteristic equations of the open and of the closed loops, it can be seen that the poles of the open loop depend only on $\boldsymbol{A}$, while the poles of the closed loop depend on three parameters $\{\boldsymbol{A}, \boldsymbol{b}, \boldsymbol{k}\}$. The performance of the closed loop is prescribed by the required location of its poles in the complex plane. What has to be found is a state feedback vector $\boldsymbol{k}$ that ensures that the roots of the characteristic equation are in the required locations.

9.1 State Feedback with Pole Placement

The design of state feedback is executed in three steps:

- choose the desired location of the poles of the closed loop system;
- for *SISO* systems the state feedback vector $\boldsymbol{k}$ can be determined by the ACKERMANN formula (textbook [1], Sect. 9.1), in MATLAB™ by using the command acker.
- determine the compensation factor k_r to fulfill the static requirements.

Example 9.1 Consider the system given by the following transfer function:

$$P(s) = \frac{6}{(s+1)(s+2)(s+3)}$$

The static gain of the system is 1, its poles are −1, −2, −3. Give the system in MATLAB™ with its poles, then transform it to state space form.

```
num=6;
den=poly([-1,-2,-3])
P=tf(num,den)
[A,b,c,d]=tf2ss(num,den)
```

The command tf2ss gives the controllability canonical form of the state equation.

$$\boldsymbol{A} = \begin{bmatrix} -6 & -11 & -6 \\ 1 & 0 & 0 \\ 0 & 1 & 0 \end{bmatrix}, \ \boldsymbol{b} = \begin{bmatrix} 1 \\ 0 \\ 0 \end{bmatrix}, \ \boldsymbol{c}^{\mathrm{T}} = \begin{bmatrix} 0 & 0 & 6 \end{bmatrix}, \ d = 0$$

Choose the $\boldsymbol{p}_{\mathrm{k}}$ poles of the closed loop by

```
pk=[-6;-3+i*4;-3-i*4]
```

(The conjugate complex poles can be considered as the poles of a second order oscillating system. The damping factor ξ is calculated from the angle φ of the vector of the poles, $\xi = \cos\varphi = 3/\sqrt{9+16} = 0.6$).

Let us remark that the system can be accelerated by shifting its poles to the left in the complex plane. Analyse the required behaviour of the step response with these prescribed poles.

First let the numerator be the constant 1, and let the denominator be the characteristic polynomial.

```
numk=1
denk=poly(pk)
H=tf(numk,denk)
H=zpk(H)
g0=dcgain(H)
```

To get a system with unit gain, normalize the system by its static gain. Compare the step responses of the original and the prescribed system.

```
Hn=H/g0
step(P,'b',Hn,'r'); grid
```

In Fig. 9.2 it can be seen that with this pole prescription the system can be accelerated significantly.

Then using the Ackermann formula determine the state feedback vector that shifts the poles $\boldsymbol{p}_{\mathrm{o}}^{\mathrm{T}} = \begin{bmatrix} -1 & -2 & -3 \end{bmatrix}$ of the open loop to the required locations

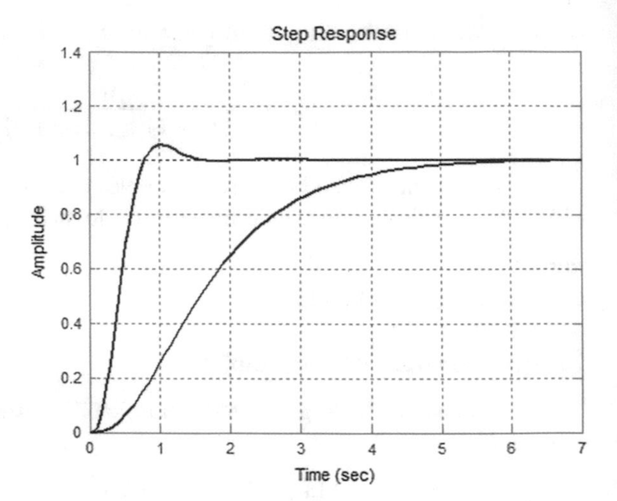

Fig. 9.2 With pole prescription the system can be accelerated

$\boldsymbol{p}_{\mathrm{k}}^{\mathrm{T}} = [-6 \quad -3+4j \quad -3-4j]$ of the closed loop. The analytical form of the ACKERMANN formula is

$$\boldsymbol{k}^{\mathrm{T}} = [1, 0, \ldots, 0]\boldsymbol{M}_{\mathrm{c}}^{-1}\mathcal{R}(\boldsymbol{A}),$$

where $\boldsymbol{M}_{\mathrm{c}}$ is the controllability matrix of the open loop, $\mathcal{R}(s)$ is the characteristic equation of the closed loop (which is determined by its prescribed poles), and $\mathcal{R}(\boldsymbol{A})$ is the value of this polynomial at $\boldsymbol{A}$. In MATLAB™ all this is executed by one command:

```
k=acker(A,b,pk)
    k = 6   50  144

Tk=ss(A-b*k,b,c,d)
Tk=zpk(Tk)
step(Tk,6)
```

In Fig. 9.3 it can be seen that by shifting the poles to the left, the transients of the step response decay faster, but the static value is not satisfactory. To ensure a static gain of value 1, a compensation factor k_{r} is calculated.

```
kr=1/dcgain(Tk)
      kr =
           25.0000
Tk1=kr*Tk
```

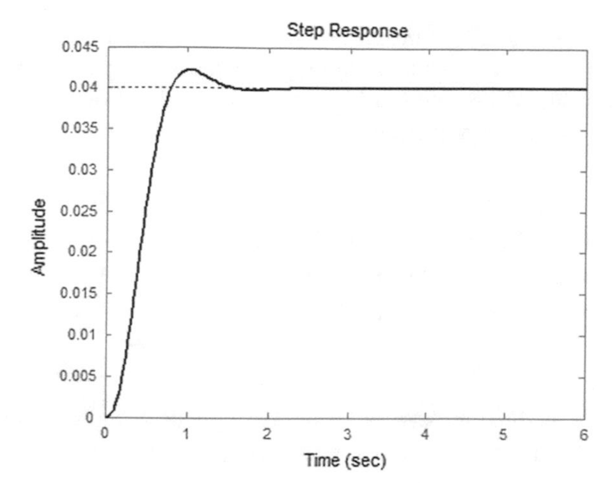

Fig. 9.3 Step response of state feedback

or

```
Tk1=ss(A-b*k,kr*b,c,d),Tk=zpk(Tk)
                    150
         ----------------------
         (s+6) (s^2  + 6s + 25)
step(Tk1,'b')
```

In Fig. 9.4 it can be seen that setting the state feedback vector $\boldsymbol{k}^{\mathrm{T}}$ and the compensation factor k_{r}, the settling process is fast and there is no static error. The dynamic properties have also been improved by this pole placement.

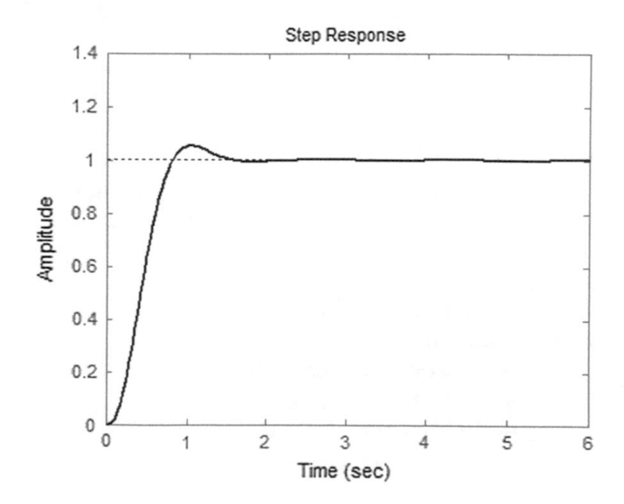

Fig. 9.4 Static error can be compensated

It should be mentioned that the choice of the state feedback vector $\boldsymbol{k}^{\mathrm{T}}$ is not unique, it depends on the form of the state space representation. Let us check the value of the state feedback vector when the state space representation of the process is given in a different form.

```
s=zpk('s')
P=6/((s+1)*(s+2)*(s+3))
[A1,b1,c1,d1]=ssdata(P)
k1=acker(A1,b1,pk)
        k1 =  40.8248  13.0639   2.4495
```

A different state representation yields a different state feedback vector. But the transfer functions of the two different representations are the same, yielding the same step responses.

```
Tk1=ss(A1-b1*k1,b1,c1,d1)
kr1= 1/dcgain(Tk1)
T1=zpk(T1)*kr1
                150
        -------------------
        (s+6) (s^2 + 6s + 25)
```

Example 9.2 With state feedback, unstable processes can be stabilized easily. The state feedback constants are calculated by prescribing stable closed loop poles.

Consider the transfer function of an unstable process containing one pole in the right half-plane:

$$P(s) = \frac{-6}{(s-1)(s+2)(s+3)}$$

Suppose that the prescribed poles of the closed loop are

```
pk=[-6;-3+i*4;-3-i*4]
```

Determine the state feedback vector and plot the step response of the closed loop. The MATLAB™ commands to do this are

```
num=-6;
den=poly([1,-2,-3])
P=tf(num,den)
[A,b,c,d]=tf2ss(num,den)
pk=[-6;-3+i*4;-3-i*4]
k=acker(A,b,pk)
```

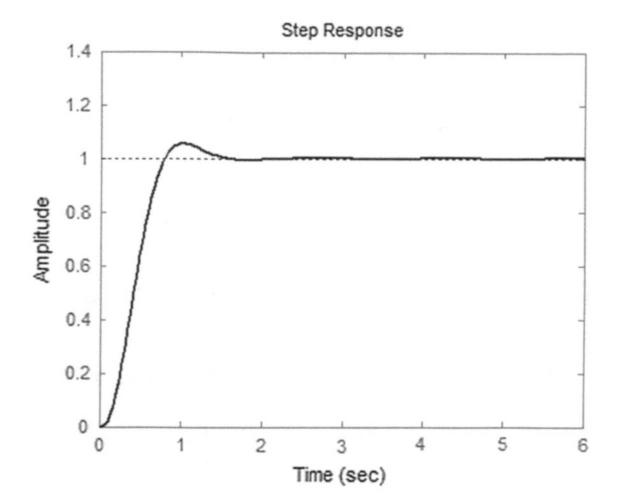

Fig. 9.5 Step response of a state feedback control system with an unstable process

```
Tk=ss(A-b*k,b,c,d)
kr=1/dcgain(Tk)
Tk1=ss(A-b*k,kr*b,c,d)
step(Tk1,6)
```

and the state feedback vector is then

```
k = 8    60    156
```

Figure 9.5 shows the step response which ensures a performance corresponding to the prescribed poles.

9.2 Introducing an Integrator into the Feedback Loop

The properties of state feedback control are analogous to the effect of serial *PD* compensation, resulting in acceleration of the control circuit. The static accuracy is ensured by a gain factor acting outside of the feedback circuit. This gain factor is determined by the knowledge of the system parameters. This means that this gain is sensitive to the accuracy of the knowledge of the parameters. Furthermore, the effect of the disturbances can not be compensated with elements outside of the feedback circuit. Therefore to ensure the static accuracy—similarly to design considerations in the frequency domain—it is expedient to introduce an integrator into the control circuit.

The state equation of the process is extended by the state variable x_i, which is the integral of the output signal y (Fig. 9.6).

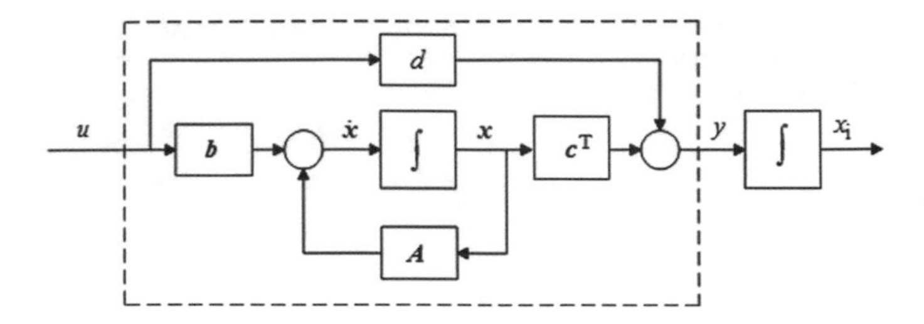

Fig. 9.6 An additional state variable is introduced as integral of the output variable

The state equation of the extended system is

$$\begin{bmatrix} \dot{\boldsymbol{x}} \\ \dot{x}_\mathrm{i} \end{bmatrix} = \begin{bmatrix} \boldsymbol{A} & 0 \\ \boldsymbol{c}^\mathrm{T} & 0 \end{bmatrix} \begin{bmatrix} \boldsymbol{x} \\ x_\mathrm{i} \end{bmatrix} + \begin{bmatrix} \boldsymbol{b} \\ 0 \end{bmatrix} u = \boldsymbol{A}_\mathrm{b} \boldsymbol{x}_\mathrm{b} + \boldsymbol{b}_\mathrm{b} u$$

$$y = \begin{bmatrix} \boldsymbol{c}^\mathrm{T} & 0 \end{bmatrix} \begin{bmatrix} \boldsymbol{x} \\ x_\mathrm{i} \end{bmatrix} + du = c_\mathrm{b}^\mathrm{T} \boldsymbol{x}_\mathrm{b} + du$$

So the number of the state variables is increased by 1. For state feedback design, the number of the prescribed poles should also increase by 1. The state feedback vector $\boldsymbol{k}_\mathrm{b}^\mathrm{T}$ is calculated now for the extended state equation with state matrices $\boldsymbol{A}_\mathrm{b}$ and $\boldsymbol{b}_\mathrm{b}$, for the prescribed poles $\boldsymbol{p}_\mathrm{b}$, using the Ackermann formula. These poles will be the prescribed poles of the characteristic equation $\det\left(s\boldsymbol{I} - \boldsymbol{A}_\mathrm{b} + \boldsymbol{b}_\mathrm{b}\boldsymbol{k}_\mathrm{b}^\mathrm{T}\right) = 0$.

Figure 9.7 shows the extended state feedback system. The integrator is located after the error signal.

Supposing a single input–single output *SISO* system and $d = 0$ the state equation of the closed loop system is written as

$$\dot{\boldsymbol{x}}_\mathrm{z} = \begin{bmatrix} \dot{\boldsymbol{x}} \\ \dot{x}_\mathrm{i} \end{bmatrix} = \begin{bmatrix} \boldsymbol{A} - \boldsymbol{b}\boldsymbol{k}^\mathrm{T} & \boldsymbol{b}k_\mathrm{i} \\ -\boldsymbol{c}^\mathrm{T} & 0 \end{bmatrix} \begin{bmatrix} \boldsymbol{x} \\ x_\mathrm{i} \end{bmatrix} + \begin{bmatrix} 0 \\ 1 \end{bmatrix} r = \boldsymbol{A}_\mathrm{z} \boldsymbol{x}_\mathrm{z} + \boldsymbol{b}_\mathrm{z} r$$

$$y = \begin{bmatrix} \boldsymbol{c}^\mathrm{T} & 0 \end{bmatrix} \begin{bmatrix} \boldsymbol{x} \\ x_\mathrm{i} \end{bmatrix} + 0 \cdot r = \boldsymbol{c}_\mathrm{z}^\mathrm{T} \boldsymbol{x}_\mathrm{z} + 0 \cdot r$$

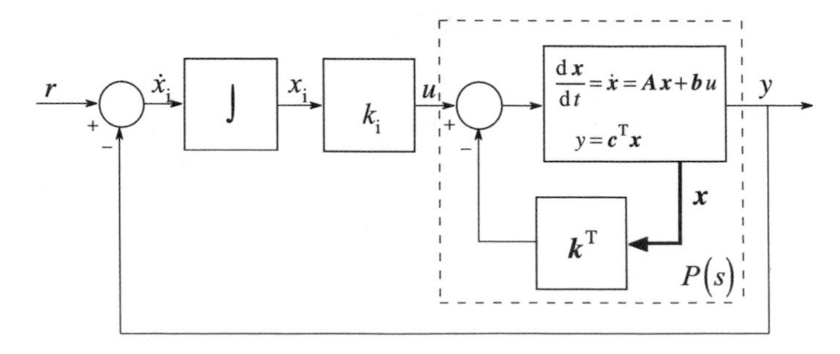

Fig. 9.7 Block diagram of the extended state feedback system

Example 9.3 Extend the process given in Example 9.1 with an integrating state variable.

```
num =6;
den =poly([-1,-2,-3])
P=tf(num,den)
[A,b,c,d]=tf2ss(num,den)
```

The parameter matrices of the extended system are

```
nulvec=[0;0;0];
Ab=[A nulvec;c 0]
bb=[b;0]
    Ab =
         -6     -11     -6     0
          1       0      0     0
          0       1      0     0
          0       0      6     0
    bb =
          1
          0
          0
          0
```

Let the poles of the closed loop system be

```
pb=[-9 -6 -3+i*4 -3-i*4];
```

Determine the state feedback vector:

```
kb=acker(Ab,bb,pb)
    kb =
         15    158    693    225
```

The first three elements of the extended state feedback vector realize the state feedback from the original state variables, while the fourth element, k_i, belongs to the artificially introduced integrator:

```
k=kb(1:3)
ki=kb(4)
    k =
```

```
     15    158    693
```

The state matrices of the closed loop system are

```
Az=[A-b*k b*ki;-c 0]
bz=[nulvec;1]
cz=[c 0]
dz=0;
   Az =
       -21  -169     -699     225
         1     0        0       0
         0     1        0       0
         0     0       -6       0
   bz =
         0
         0
         0
         1
   cz =
         0     0        6       0
   dz=0
```

The step response of the closed loop (Fig. 9.8) is found by

```
t=0:0.1:6;
step(Az,bz,cz,dz,1,t),grid
```

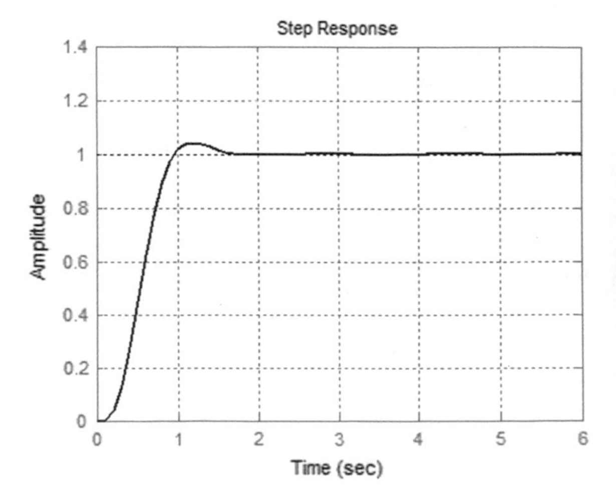

Fig. 9.8 Step response of the closed loop

It can be seen that the dynamic and static behaviour of the closed loop system is appropriate.

9.3 State Estimation

In practical applications, the instrumentation of the processes includes possibilities for measurement of several variables. Sensors measure the output signal, but generally not all the state variables are available for measurement. In this case the control with state feedback has to be supplemented with the estimation of the non-measurable state variables. The block scheme of a *state estimator* is shown in Fig. 9.9. The estimator contains the model of the system. It is assumed that $d = 0$. If the system is known, the parameter matrices of the model are the same as the parameter matrices of the system. The difference between the output of the system and the model constitutes an error signal. This error signal is fed back to the summing point at the derivatives of the estimated variables to modify their values. The aim is to ensure that the estimated state variables move quickly to follow the movement of the real state variables. The state estimation circuit forms a closed loop whose input signal is y, the output signal of the process. The poles of the estimation circuit can be prescribed. An important requirement is that the dynamics of the estimation circuit should be much faster than the dynamics of the process. The gain $\boldsymbol{l}$ of the estimation circuit can be calculated by the ACKERMANN formula. It can be seen in the figure that the behaviour of the estimation circuit is influenced by

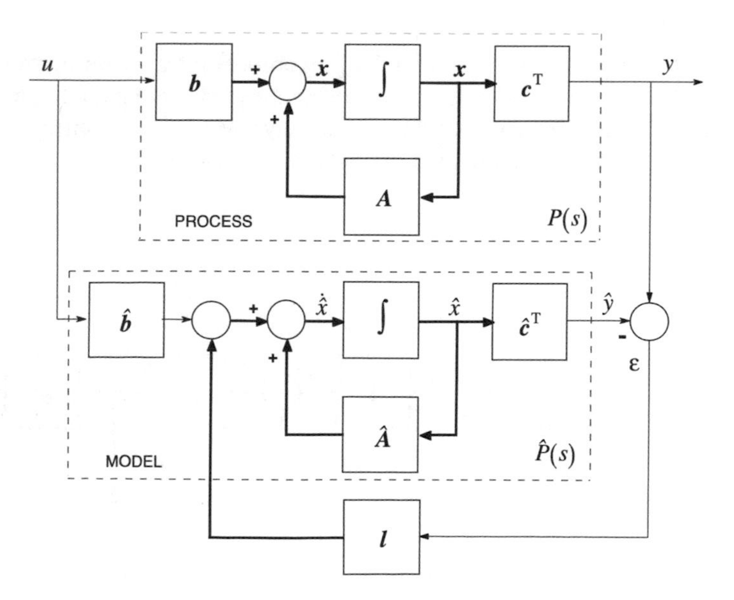

Fig. 9.9 Block diagram of state estimation

the parameter matrices $\hat{\boldsymbol{A}}$ and $\hat{\boldsymbol{c}}^{\mathrm{T}}$. (For simplicity, $\boldsymbol{A}$, $\boldsymbol{b}$ and $\boldsymbol{c}^{\mathrm{T}}$ are used in the formulas.)

Let us suppose that the parameter matrices of the process and of the model are the same ($\varepsilon = 0$). The free motion of the system states is to be estimated, i.e. the motion of the state variables starting from their initial values supposing a zero input signal. The output disturbance is zero. Based on Fig. 9.9, the estimated state variables can be calculated according to the following relation:

$$\dot{\hat{\boldsymbol{x}}} = \boldsymbol{A}\hat{\boldsymbol{x}} + \boldsymbol{b}u + \boldsymbol{l}\boldsymbol{c}^{\mathrm{T}}(\boldsymbol{x} - \hat{\boldsymbol{x}}) = \left(\boldsymbol{A} - \boldsymbol{l}\boldsymbol{c}^{\mathrm{T}}\right)\hat{\boldsymbol{x}} + \boldsymbol{l}y.$$

Let us introduce the error signal $\boldsymbol{\varepsilon} = \boldsymbol{x} - \hat{\boldsymbol{x}}$. The derivative of the error signal is obtained if the equation given for the estimated state variables is subtracted from the equation of the original state variables.

$$\dot{\boldsymbol{x}} - \dot{\hat{\boldsymbol{x}}} = \dot{\varepsilon} = \boldsymbol{A}\,\varepsilon - \boldsymbol{l}\boldsymbol{c}^{\mathrm{T}}\varepsilon = \left(\boldsymbol{A} - \boldsymbol{l}\boldsymbol{c}^{T}\right)\varepsilon = \boldsymbol{A}_{\varepsilon}\varepsilon = \boldsymbol{A}\,\varepsilon - \boldsymbol{l}(y - \hat{y})$$

The estimation circuit can be redrawn as Fig. 9.10.

The parameters of the estimation circuit (the elements of the vector $\boldsymbol{l}$) can be calculated by the Ackermann formula prescribing the roots of the characteristic equation of the closed estimation circuit.

```
L=acker(A',c',Pe)'
```

Here Pe is the vector of the prescribed poles of the estimation circuit. The estimation circuit has to be faster than the process, and faster than the control system with state feedback. (Transposition is required to reconcile the dimensions of the matrices and the vectors.)

Example 9.4 The process is the third order proportional system investigated also in Example 9.1 (without the extension by the integrating state variable). Let the initial conditions of all the three state variables have the value 1. The reference signal and the disturbance signal are zero. Give the poles of the estimation circuit as

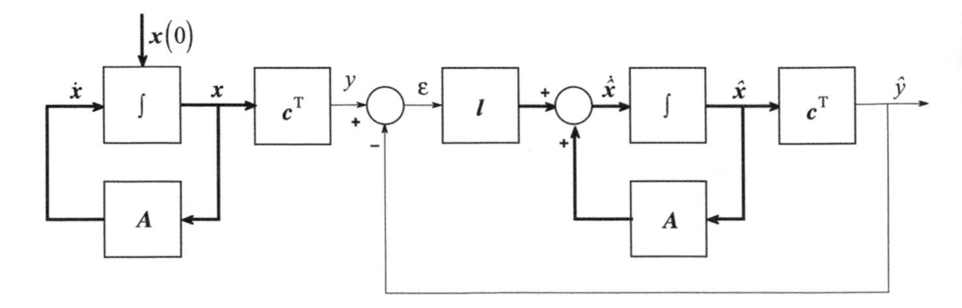

Fig. 9.10 The redrawn estimation circuit

```
Pe=[-7    -7    -7]
```

The state estimation vector is obtained as $l^{\mathrm{T}} = [-17.3333 \quad 7.6667 \quad 2.5000]$. The MATLAB™ program below gives the course in time of the real state variables of the system which have to be estimated, then calculates the vector $\boldsymbol{l}$ of the estimation circuit. Then according to Fig. 9.10 it simulates the evolution in time of the state estimation exciting the estimation circuit with the signal y as the input of the circuit. The program plots in one diagram the real state variables and their estimation, as well as the output signal and its estimated value.

```
clear
clc
num =6;
den =poly([-1,-2,-3])
P=tf(num,den)
[A,b,c,d]=tf2ss(num,den)
sys1=ss(A,b,c,d)
x0=[1;1;1]
t=0:0.05:6;
[y,t,x]=initial(sys1,x0,t);
figure(1)
plot(t,x),grid
Pe=[-7 -7 -7]
L=acker(A',c',Pe)
Aest=A-L'*c
sysest=ss(Aest,L',c,d)
x0est=[0;0;0]
[yest,t,xest]=lsim(sysest,y,t,x0est)
figure(2)
plot(t,x,t,xest),grid
figure(3)
plot(t,x(:,1),t,xest(:,1)),grid
figure(4)
plot(t,y,t,yest),grid
```

Plot the evolution in time of the first state variable and its estimated value (Fig. 9.11). The simulation shows that the state variables become settled quickly.

```
plot(t,[x(:,1),xest(:,1)]),grid
```

Prescribing appropriate poles of the estimation circuit, the settling process can be further accelerated and the transients of the estimation can be influenced.

Build the state estimation circuit also in SIMULINK™. The process and its model are built from the blocks *State-Space* of the *Continuous* library, and the *Matrix Gain* block of the library *Math Operations*. The separation of parameter $\boldsymbol{c}$ is needed because not only the output signal, but also the state variables have to be

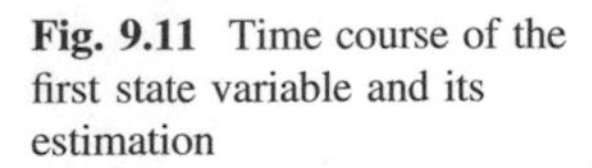

Fig. 9.11 Time course of the first state variable and its estimation

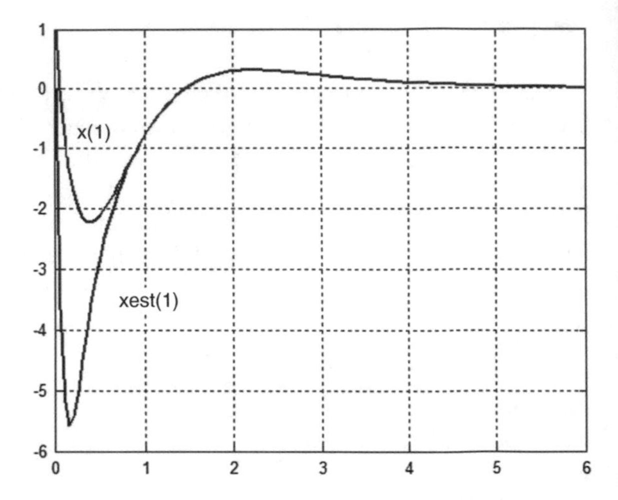

reached. The parameter $\boldsymbol{b}$ is also separated from the state model block, as the derivatives of the state variables are modified, so the derivatives have to be also available. (So in the *State-Space* blocks in the SIMULINK™ model (Fig. 9.12), the parameters $\boldsymbol{B}$ and $\boldsymbol{C}$ are the identity matrices of the appropriate dimensions, and the parameter d is a zero matrix. The process and its model can be the same, if the process is known.) In the SIMULINK™ diagram shown in Fig. 9.12, the changes in the real and the estimated state variables can be followed not only as the effect of the unknown initial conditions, but also for the input and the disturbance signals. In the example, after determining the state equation of the process and the calculation of the vector $\boldsymbol{l}$ of the estimation circuit, the SIMULINK™ block can be run. In the figure the parameters set for the *State-Space* blocks and the *Matrix Gain* blocks are shown. Running the program, it can be observed in *Scope* that the estimated state variables quickly follow the real state variables. As the variables are connected also to *Workspace* blocks, the real and the estimated state variables can be plotted from the MATLAB™ surface as well. For the course in time of the first state variable and of its estimation, the result is the same as given in Fig. 9.11.

```
plot(t,x,t,xest),grid
```

Problem Set the values of the initial conditions to zero and the value of the output disturbance to 1. Running the simulation, it can be seen that there is a static deviation between the real and the estimated state variables. The input signal excites the real and the estimation circuit the same way, therefore this excitation will not distort the state estimation. But the output disturbance excites them differently, and therefore a static error will appear in the estimation. To eliminate this deviation the disturbance signal should be described by its state variables, then the state equation should be enhanced by the state variables of the disturbance. Then the state estimation could be executed for the extended system (but this extension will no be not dealt with in more detail here).

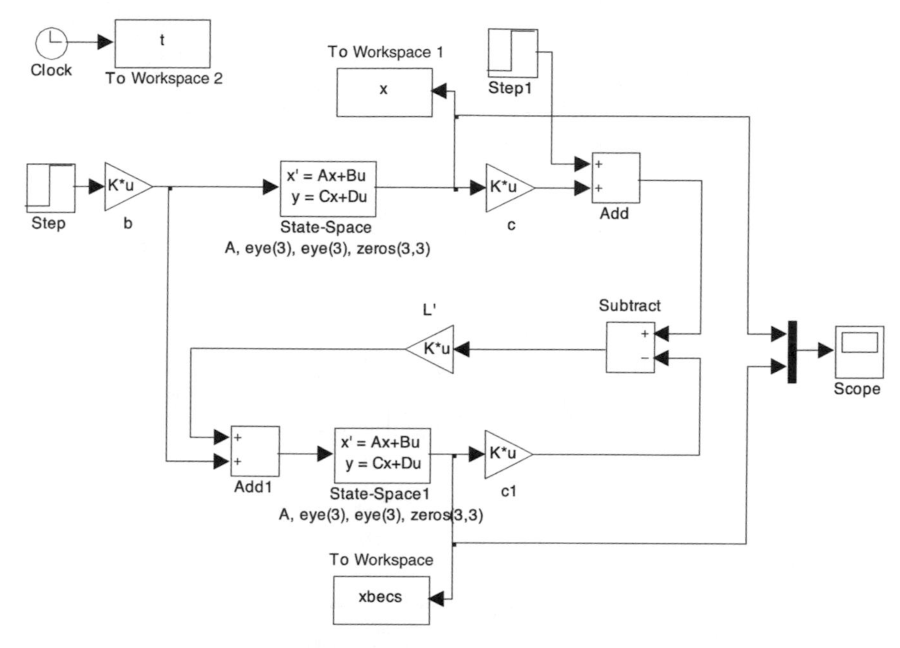

Fig. 9.12 SIMULINK™ diagram of state estimation

9.4 State Feedback with State Estimation

State estimation (observer) and state feedback can be executed independently of each other (separation principle, textbook [1], Chap. 9). If the state variables are not available, then state feedback control can be realized by feeding back the estimated state variables with the state feedback vector $\boldsymbol{k}$ calculated for the original state variables. An important principle is that the dynamics of the state feedback circuit should be faster than the process dynamics, and the dynamics of the estimation circuit should be faster than the dynamics of the state feedback circuit to ensure that the state feedback would consider estimated state variables which approach quickly and well the state variables of the real system.

The block diagram of the state feedback system using an observer is given in Fig. 9.13. On the basis of the figure, the state equation of the system is

$$\begin{bmatrix} \dot{\boldsymbol{x}} \\ \dot{\hat{\boldsymbol{x}}} \end{bmatrix} = \begin{bmatrix} \boldsymbol{A} & -\boldsymbol{b}\boldsymbol{k}^{\mathrm{T}} \\ \boldsymbol{l}\boldsymbol{c}^{\mathrm{T}} & \boldsymbol{A} - \boldsymbol{l}\boldsymbol{c}^{\mathrm{T}} - \boldsymbol{b}\boldsymbol{k}^{\mathrm{T}} \end{bmatrix} \begin{bmatrix} \boldsymbol{x} \\ \hat{\boldsymbol{x}} \end{bmatrix} + \begin{bmatrix} \boldsymbol{b}k_{\mathrm{r}} \\ \boldsymbol{b}k_{\mathrm{r}} \end{bmatrix} r$$

$$y = \begin{bmatrix} \boldsymbol{c}^{\mathrm{T}} & 0 \end{bmatrix} \begin{bmatrix} \boldsymbol{x} \\ \hat{\boldsymbol{x}} \end{bmatrix} + 0 \cdot r$$

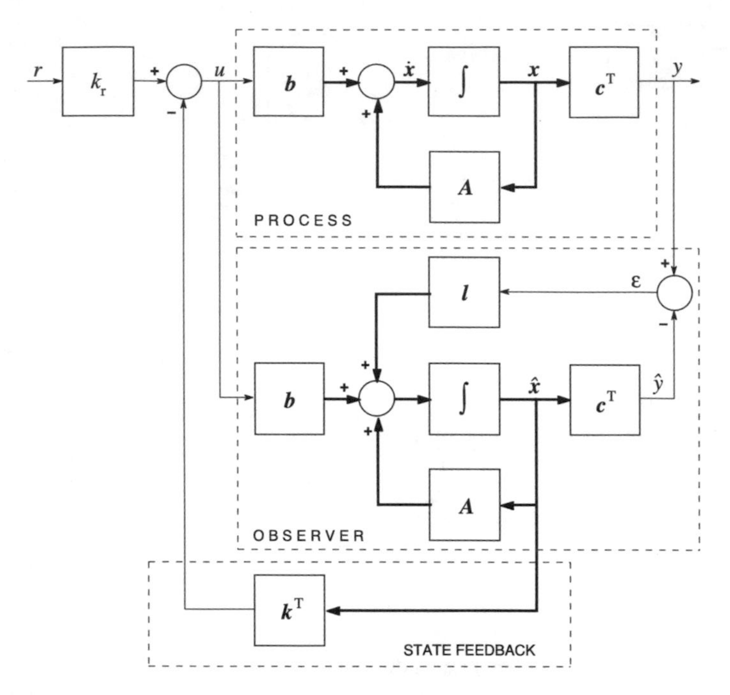

Fig. 9.13 State feedback from the estimated state variables

Example 9.5 Let us approximate the state variables of Example 9.1 by the estimated state variables calculated in Example 9.4, then feed back the approximate state variables by the state feedback constants calculated in Example 9.1. Plot in one diagram the step response of the output signals for the case when the feedback is taken from the original, and for the case where it is taken from the estimated state variables. The reference signal is a unit step and the initial conditions are $[0.2 \quad 0.2 \quad 0.2]$. The initial conditions of the observer states are supposed to be zeros.

The MATLAB™ program is

```
clear;clc
num =6;
den =poly([-1,-2,-3])
P=tf(num,den)
[A,b,c,d]=tf2ss(num,den)
sys1=ss(A,b,c,d)
%process
pk=[-6;-3+i*4;-3-i*4]
k=acker(A,b,pk)
```

```
sys2=ss(A-b*k,b,c,d)
kr=1/dcgain(sys2)
sys3=ss(A-b*k,b*kr,c,d)
%state feedback system
x0=[0.2;0.2;0.2]
t=0:0.05:3;
[y1,t,x] = initial(sys3,x0,t);
y2=step(sys3,t);
y=y1+y2;
%Output of the state feedback system
pe=[-7 -7 -7]
L=acker(A',c',pe)
Abvcs=[A -b*k;L'*c A-L'*c-b*k]
bbvcs=[b*kr;b*kr]
cbvcs=[c zeros(1,3)]
dbvcs=0
sys4=ss(Abvcs,bbvcs,cbvcs,dbvcs)
x0est=[0;0;0]
x0bvcs=[x0;x0est]
[y3,t,x3] = initial(sys4,x0bvcs,t);
y4=step(sys4,t);
y5=y3+y4;
%Output of the state feedback system from the estimated
%state variables
figure(1)
plot(t,y,t,y5),grid
figure(2)
plot(t,x3),grid
%The real and the estimated state variables
figure(3)
plot(t,x3(:,1),t,x3(:,4)),grid
figure(4)
plot(t,x3(:,2),t,x3(:,5)),grid
figure(5)
plot(t,x3(:,3),t,x3(:,6)),grid
```

Figure 9.14 gives the output signals of the state feedback systems when the feedback is taken from the original, and when it is taken from the estimated state variables. The overshoot is higher in the case when the feedback is taken from the estimated state variables. Figure 9.15 shows the first state variable and its estimation.

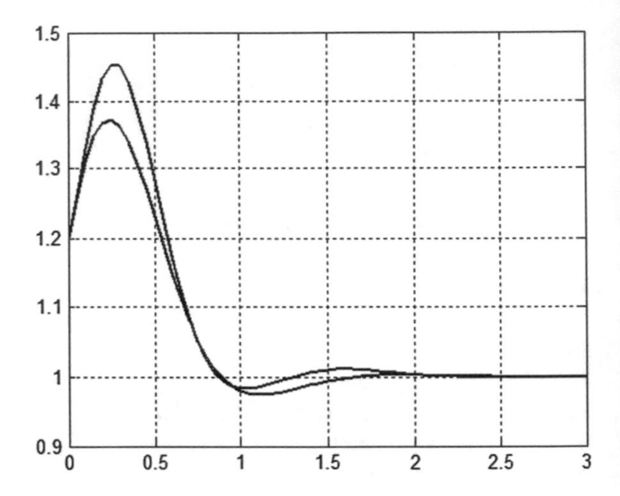

Fig. 9.14 Output signals with feedback from the real and from the estimated state variables

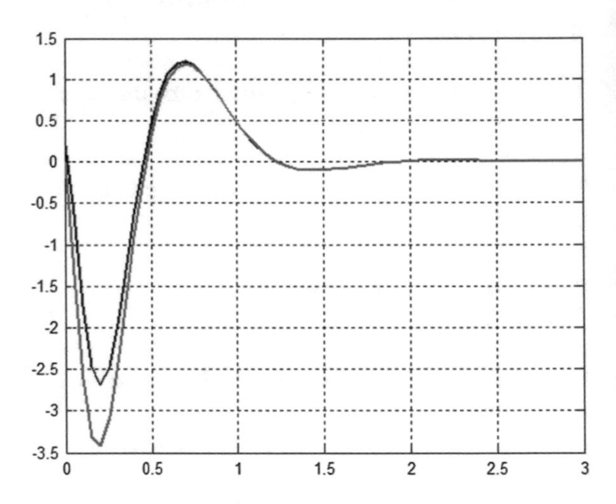

Fig. 9.15 The time course of the first state variable and its estimation

Figure 9.16 gives the SIMULINK™ block diagram of the control system. Running it for the given initial conditions and for unit step reference signal the obtained results coincide with the results obtained in MATLAB™. With the output disturbance, a static error does appear in the estimation of the state variables and also in the output signal.

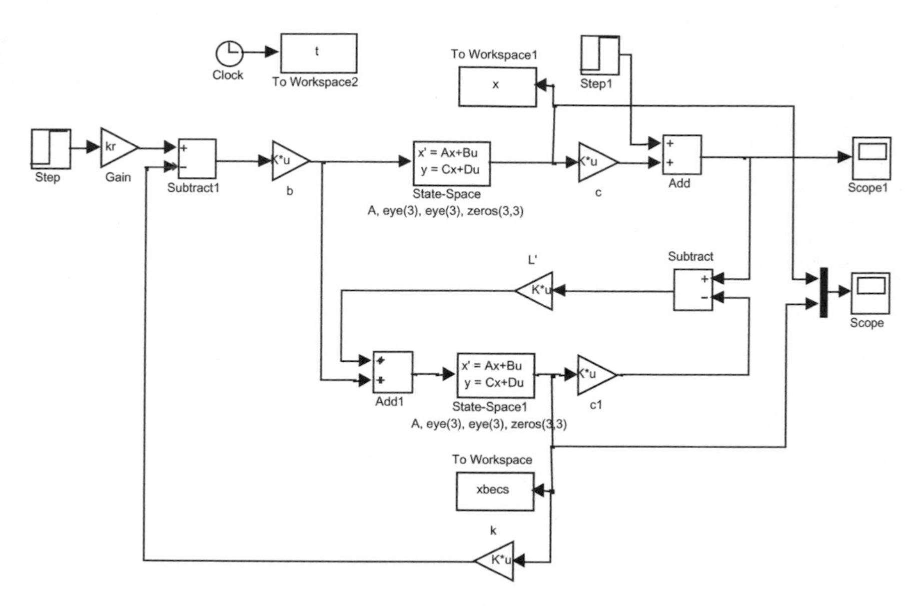

Fig. 9.16 SIMULINK™ diagram of state feedback taken from the estimated state variables

Problem Build a SIMULINK™ block diagram when extending the system with the integrating state variable with state estimation and state feedback from the estimated state variables. Simulate the behaviour of the control system for unit step reference signal with the initial conditions given before. Analyse the disturbance rejection properties of the system in the case of an output disturbance.

Chapter 10
General Polynomial Method for Regulator Design

The theoretical considerations of Chap. 10 of the textbook [1] are summarized here to support the introduction of the variable names in the corresponding MATLAB™ programs.

<u>Process</u>: $P = \frac{\mathcal{B}}{\mathcal{A}}$

<u>Regulator</u>: $C = \frac{\mathcal{Y}}{\mathcal{X}}$

<u>Resulting transfer function</u>: $T = \frac{CP}{1+CP} = \frac{\frac{\mathcal{Y}\mathcal{B}}{\mathcal{X}\mathcal{A}}}{1+\frac{\mathcal{Y}\mathcal{B}}{\mathcal{X}\mathcal{A}}} = \frac{\mathcal{Y}\mathcal{B}}{\mathcal{Y}\mathcal{B}+\mathcal{X}\mathcal{A}} = \frac{\mathcal{Y}\mathcal{B}}{\mathcal{R}}$, where the characteristic polynomial of the closed loop is $\mathcal{R}(s) = \mathcal{X}(s)\mathcal{A}(s) + \mathcal{Y}(s)\mathcal{B}(s)$ and the characteristic equation is $\mathcal{R}(s) = \mathcal{X}(s)\mathcal{A}(s) + \mathcal{Y}(s)\mathcal{B}(s) = 0$.

<u>Sensitivity function</u>: $S = \frac{1}{1+CP} = \frac{1}{1+\frac{\mathcal{Y}\mathcal{B}}{\mathcal{X}\mathcal{A}}} = \frac{\mathcal{X}\mathcal{A}}{\mathcal{Y}\mathcal{B}+\mathcal{X}\mathcal{A}} = \frac{\mathcal{X}\mathcal{A}}{\mathcal{R}}$

Let the order of the system be n, i.e. $\deg\{\mathcal{A}\} = n$. A realizable regulator $C(s)$ is to be given, which

- yields the given characteristic polynomial $\mathcal{R}(s)$. The regulator is determined by solving the Diophantine equation $\mathcal{X}(s)\mathcal{A}(s) + \mathcal{Y}(s)\mathcal{B}(s) = \mathcal{R}(s)$
- at the initial time instant $t = 0$ for a unit step reference signal it provides a control signal value $u(0) \neq 0$, i.e. $\deg\{\mathcal{X}\} = \deg\{\mathcal{Y}\}$.

An important remark is that in the resulting transfer function of the closed-loop $T(s) = \frac{\mathcal{Y}(s)\mathcal{B}(s)}{\mathcal{R}(s)}$

- $\mathcal{R}(s)$ is a <u>given</u> polynomial determined by the designer.
- $\mathcal{Y}(s)$ <u>is calculated</u> by solving the Diophantine equation $\mathcal{X}(s)\mathcal{A}(s) + \mathcal{Y}(s)\mathcal{B}(s) = \mathcal{R}(s)$.
- $\mathcal{B}(s)$ <u>is given</u>: it is the numerator of the transfer function of the process.

Let us choose the degree of the characteristic polynomial $\mathcal{X}(s)\mathcal{A}(s) + \mathcal{Y}(s)\mathcal{B}(s) = \mathcal{R}(s)$ to be $\deg\{\mathcal{R}\} = 2n - 1$. Then the Diophantine equation always has a solution and the degree of the regulator is $(n - 1)$.

L. Keviczky et al., *Control Engineering: MATLAB Exercises*, Advanced Textbooks in Control and Signal Processing,
https://doi.org/10.1007/978-981-10-8321-1_10

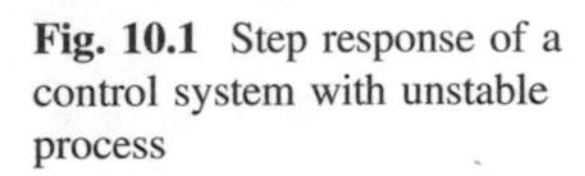

Fig. 10.1 Step response of a control system with unstable process

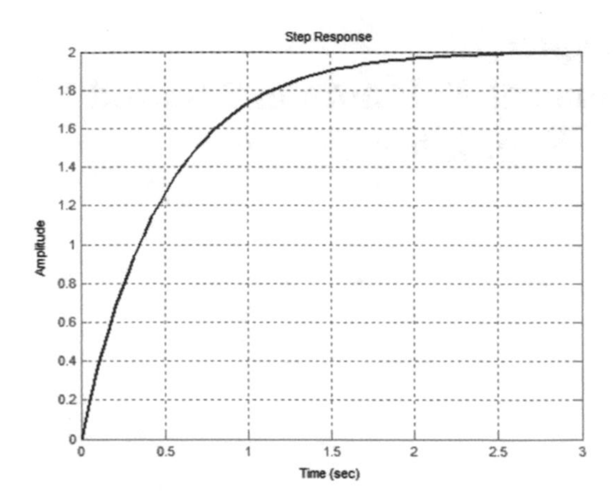

Example 10.1 Repeating Example 10.1 of the textbook [1] the transfer function of the unstable process is $P(s) = \frac{-1}{s-2}$. Then $n = 1$ and the degree of the regulator is 0, $C(s) = \frac{\mathcal{Y}}{\mathcal{X}} = \frac{K}{1} = K$. Let the first order characteristic polynomial be $\mathcal{R}(s) = s+2$ (The unstable pole of the process is reflected in the imaginary axis). Then from the solution of the DIOPHANTINE equation $\mathcal{X}(s)\mathcal{A}(s) + \mathcal{Y}(s)\mathcal{B}(s) = 1\,(s-2) + K(-1) = \mathcal{R}(s) = s+2$, the gain $K = -4$ is obtained for the regulator.

```
s=zpk('s')
P=-1/(s-2)
C=-4
T=C*P/(1+C*P), T=minreal(T)
          4
        -----
        s + 2
step(T),grid
```

In Fig. 10.1 it can be seen that with the designed regulator the unstable process has been stabilized, the characteristic equation of the closed loop is indeed $\mathcal{R}(s) = s+2$. But there is a significant static error: $\lim_{t\to\infty} y(t) = 2$. With a constant $F(s) = 0.5$ precompensator the static error becomes zero (Fig. 10.2).

```
step(0.5*T),grid
```

Let us remark that the resulting transfer function of the closed loop with the precompensator is $T_e(s) = F(s)\frac{\mathcal{Y}(s)\mathcal{B}(s)}{\mathcal{R}(s)}$. With the precompensator $F(s)$, the effect of the zeros in the numerator of $T(s)$ can also be compensated. If the transfer function

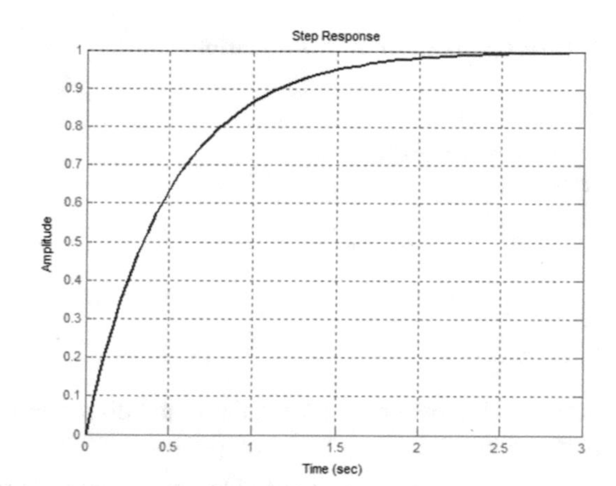

Fig. 10.2 The precompensator ensures accurate settling

$F(s)$ is not a constant, the dynamics of reference signal tracking and the dynamics of output disturbance rejection will be different, the control system will become of *two-degree-of-freedom* (2DOF).

Example 10.2 Take into consideration that the stable poles of the process (which are included in the polynomial $\mathcal{A}_+$) and its inverse stable zeros (included in the polynomial $\mathcal{B}_+$) can be cancelled with the regulator. ($\mathcal{A}_-$ and $\mathcal{B}_-$ denote the non-cancellable factors.) The transfer function of the process can be given by $P = \frac{\mathcal{B}_+}{\mathcal{A}_+}\frac{\mathcal{B}_-}{\mathcal{A}_-}$, and the regulator transfer function is expressed as $C = \frac{\mathcal{A}_+}{\mathcal{B}_+}\frac{\mathcal{Y}}{\mathcal{A}\mathcal{X}}$.

Following the notations of the textbook [1] the polynomials are factored as $P(s) = P_+(s)P_-(s)$ where the roots of $P_+(s)$ are located in the left half-plane.

The resulting transfer function of the closed loop is

$$T = \frac{CP}{1+CP} = \frac{\frac{\mathcal{A}_+}{\mathcal{B}_+}\frac{\mathcal{Y}}{\mathcal{X}}\frac{\mathcal{B}_+}{\mathcal{A}_+}\frac{\mathcal{B}_-}{\mathcal{A}_-}}{1+\frac{\mathcal{A}_+}{\mathcal{B}_+}\frac{\mathcal{Y}}{\mathcal{X}}\frac{\mathcal{B}_+}{\mathcal{A}_+}\frac{\mathcal{B}_-}{\mathcal{A}_-}} = \frac{\frac{\mathcal{Y}}{\mathcal{X}}\frac{\mathcal{B}_-}{\mathcal{A}_-}}{1+\frac{\mathcal{Y}}{\mathcal{X}}\frac{\mathcal{B}_-}{\mathcal{A}_-}} = \frac{\mathcal{Y}\mathcal{B}_-}{\mathcal{X}\mathcal{A}_- + \mathcal{Y}\mathcal{B}_-} = \frac{\mathcal{Y}\mathcal{B}_-}{\mathcal{R}},$$

where $\mathcal{X}\mathcal{A}_- + \mathcal{Y}\mathcal{B}_- = \mathcal{R}$. The sensitivity function is

$$S = \frac{1}{1+CP} = \frac{1}{1+\frac{\mathcal{A}_+}{\mathcal{B}_+}\frac{\mathcal{Y}}{\mathcal{X}}\frac{\mathcal{B}_+}{\mathcal{A}_+}\frac{\mathcal{B}_-}{\mathcal{A}_-}} = \frac{1}{1+\frac{\mathcal{Y}}{\mathcal{X}}\frac{\mathcal{B}_-}{\mathcal{A}_-}} = \frac{\mathcal{X}\mathcal{A}_-}{\mathcal{X}\mathcal{A}_- + \mathcal{Y}\mathcal{B}_-} = \frac{\mathcal{X}\mathcal{A}_-}{\mathcal{R}} = 1 - T.$$

If the transfer function of the process is $P(s) = \frac{s+7}{(s-2)(s+10)}$, then $\mathcal{B}_+ = s+7$, $\mathcal{B}_- = 1$, $\mathcal{A}_+ = s+10$ and $\mathcal{A}_- = s-2$. The DIOPHANTINE equation $\mathcal{X}\mathcal{A}_- + \mathcal{Y}\mathcal{B}_- = \mathcal{R}$ with $\frac{\mathcal{Y}}{\mathcal{X}} = \frac{K}{1}$ can be of first degree. Let us choose $\mathcal{R}(s) = s+2$ as the characteristic polynomial. So $\mathcal{X}\mathcal{A}_- + \mathcal{Y}\mathcal{B}_- = \mathcal{R}$ and $K = 4$. The regulator is $C = \frac{\mathcal{A}_+}{\mathcal{B}_+}\frac{\mathcal{Y}}{\mathcal{X}} = \frac{s+10}{s+7}K = 4\,\frac{s+10}{s+7}$.

Steps of the MATLAB™ simulation:

```
C=4*(s+10)/(s+7)
P=(s+7)/(s-2)/(s+10)
T=C*P/(1+C*P)
T=minreal(T)
        4
      -----
      s + 2
step(T),grid
```

Figure 10.3 shows that the regulator stabilizes the unstable process. The static error can be eliminated by a precompensator.

Example 10.3 Consider the plant given by the transfer function $P(s) = \frac{(s-5)(s+7)}{(s-2)(s+10)}$. Here $\mathcal{B}_+ = s+7$, $\mathcal{B}_- = s-5$, $\mathcal{A}_+ = s+10$ and $\mathcal{A}_- = s-2$. Let one root of the characteristic polynomial be again $s = -2$, and the other $s = -6$. The characteristic polynomial is now $\mathcal{R}(s) = K(s+2)(s+6)$, so $\frac{\mathcal{Y}}{\mathcal{X}} = \frac{s-z}{s-p}$ and the characteristic equation can be written as $\mathcal{X}\mathcal{A}_- + \mathcal{Y}\mathcal{B}_- = \mathcal{R}$

$$(s-p)(s-2) + (s-z)(s-5) = \mathcal{R}(s) = K(s+2)(s+6)$$

Comparing the coefficients $K = 2$, $z = 70/3$, $p = -139/3$; and the regulator is $C = \frac{\mathcal{A}_+}{\mathcal{B}_+}\frac{\mathcal{Y}}{\mathcal{X}} = \frac{s+10}{s+7}\frac{s-70/3}{s+139/3}$.

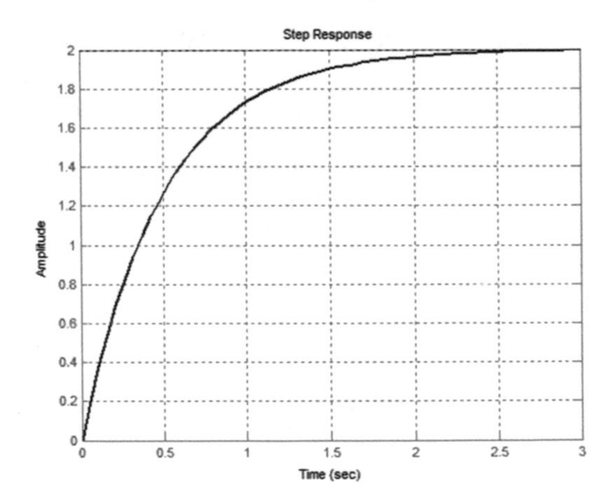

Fig. 10.3 Step response of a control system with unstable process

Steps of the simulation in MATLAB™:

```
s=zpk('s')
P=(s-5)*(s+7)/(s-2)/(s+10)
C=(s+10)*(s-70/3)/(s+7)/(s+139/3)
T=C*P/(1+C*P)
T=minreal(T)
   0.5(s-5)(s-23.33)
  ---------------
      (s+6)(s+2)
figure(1)
step(T),grid
```

The step response of the control system is shown in Fig. 10.4. If we would like to eliminate the static error and decrease the under sweeping that resulted because of the non-minimum phase feature of the process, the precompensator can be extended by a filter allocating a pole e.g. to $s = -1$: $F(s) = \frac{1}{(s+1)T(0)}$, where $T(0)$ is the static gain of the system without the precompensator. The MATLAB™ code for this is

```
F= 1/(dcgain(T)*(s+1))
figure(2)
step(F*T,6),grid
```

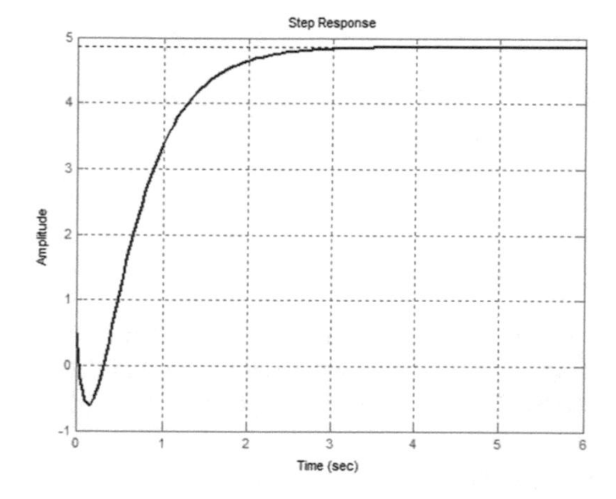

Fig. 10.4 Step response of a control system with unstable and non-minimumphase process

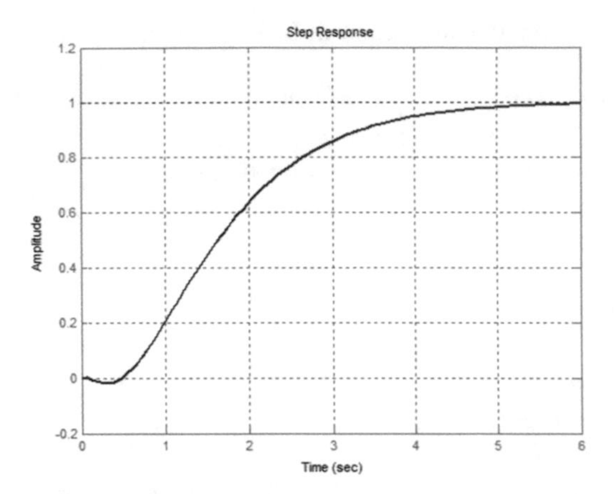

Fig. 10.5 Step response of a control system with unstable and non-minimumphase process with precompensator

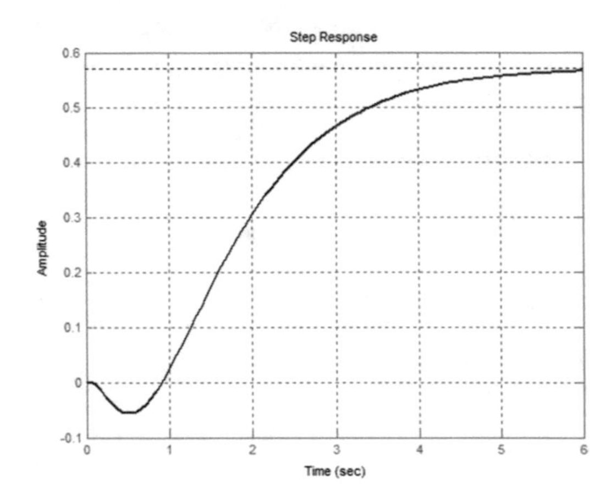

Fig. 10.6 The control signal

It can be seen that with the filter introducing the pole at $s = -1$ the undersweeping has been decreased significantly, but the settling time has been increased (Fig. 10.5). The control signal is given in Fig. 10.6.

```
U=F*C/(1+C*P)
figure(3)
step(U,6),grid
```

Refining the design of the precompensator the behaviour of the control system can be improved further.

Example 10.4 As was seen in Example 10.3, the regulator contains fixed and calculated components. The fixed components result from cancellation of the stable process poles and the inverse stable zeros; the calculated components result from the solution of the DIOPHANTINE equation. There are practical cases when it is favourable to include a further given component in the regulator. As was seen in Example 4.2, Sect. 4.4 of the textbook [1], if the requirement is to follow an exponential reference signal then a zero in the regulator corresponding to the pole of the exponential signal would ensure tracking without error. Similarly, a pole at $s = 0$ in the regulator forces an integrating effect in the regulator. Let the structure of the regulator be $C = \frac{\mathcal{A}_+}{\mathcal{B}_+}\frac{\mathcal{Y}}{\mathcal{X}}\frac{\mathcal{Y}_d}{\mathcal{X}_d}$, where $\mathcal{Y}_d$ and $\mathcal{X}_d$ are polynomials representing the given zeros and poles, respectively. Now the characteristic polynomial is $\mathcal{X}\mathcal{X}_d\mathcal{A}_- + \mathcal{Y}\mathcal{Y}_d\mathcal{B}_- = \mathcal{R}$. To ensure the solvability of the DIOPHANTINE equation considering the degrees of polynomials $\mathcal{A}_-$ and $\mathcal{B}_-$, the degrees of $\mathcal{Y}_d$, $\mathcal{X}_d$ and $\mathcal{R}$ should be chosen according to quite complex conditions.

Let us consider the process $P(s) = \frac{(s-5)(s+7)}{(s-2)(s+10)}$ analysed in Example 10.3. Here $\mathcal{B}_+ = s+7$, $\mathcal{B}_- = s-5$, $\mathcal{A}_+ = s+10$ and $\mathcal{A}_- = s-2$. Apply an integrator in the regulator, $\mathcal{X}_d = s$ and $\mathcal{Y}_d = 1$. The characteristic polynomial is $\mathcal{X}\mathcal{X}_d\mathcal{A}_- + \mathcal{Y}\mathcal{Y}_d\mathcal{B}_- = \mathcal{R}$.

In this example the essence of polynomial design is summarized in three points:

- An integrator is introduced in the loop transfer function because of the required static accuracy.
- The performance of the closed loop as the aim of the design is specified by prescribing the poles of the closed loop transfer function, i.e. prescribing the characteristic equation of the closed loop.
- The degrees of the polynomials in the numerator and the denominator of the regulator should be the same, otherwise the regulator would be non-realizable, or at the instant when the error appears it would not produce a control signal which is proportional to the value of the error.

In our example for introducing an integrator let be $\mathcal{X}_d = s$ and $\mathcal{Y}_d = 1$.

Suppose the degrees of the numerator and the denominator of the regulator are the same. $\deg\{\mathcal{A}_+\} + \deg\{\mathcal{Y}\} = \deg\{\mathcal{B}_+\} + \deg\{\mathcal{X}\} + 1$; in our case $1 + \deg\{\mathcal{Y}\} = 1 + \deg\{\mathcal{X}\} + 1$.

If the degree of $\mathcal{X}$ is zero, then the degree of $\mathcal{Y}$ is 1.

Let the prescribed roots of the characteristic equation be -2 and -6.

The characteristic equation is then

$$sx_o(s-2) + (y_o + sy_1)(s-5) = \alpha(s+2)(s+6).$$

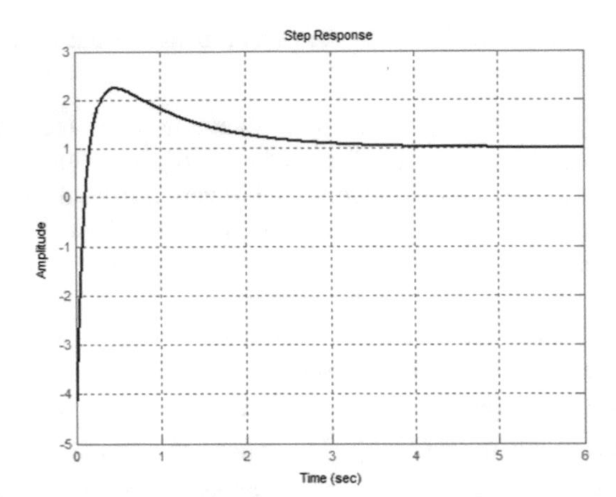

Fig. 10.7 Step response of the control system with a controller containing integrator

The solution of the DIOPHANTINE equation taking $\alpha = 1$ is $x_o = 231/31$, $y_o = -12/5$ and $y_1 = -6$. So the transfer function of the stabilizing regulator is

$$C(s) = \frac{-0.80519(s+0.4)(s+10)}{s(s+7)}$$

The step response of the control system is shown in Fig. 10.7. The dynamics and the overshoot can be modified by the polynomials $\mathcal{X}_d$ and $\mathcal{Y}_d$.

Chapter 11
Analysis of Sampled-Data Systems

11.1 Discrete-Time Systems

11.1.1 *z-Transforms*

In sampled-data systems, the signals are handled and stored in digital form. In these systems continuous-time signals are sampled periodically at sampling instants and their processing takes place in the discrete-time domain. Thus, the signals available in continuous-time control systems should be sampled and converted to digital form. Denote the sampling time by T_s. Then the digital form of a sampled continuous-time signal $y(t)$ can be represented by a train of impulses as follows:

$$\begin{aligned} y_d[k] &= y(t = kT_s) \\ &= y(0)\delta(t) + y(T_s)\delta(t - T_s) + y(2T_s)\delta(t - 2T_s) + y(3T_s)\delta(t - 3T_s) + \cdots . \end{aligned}$$

The LAPLACE transform of $y(t = kT_s) = y_d[k]$ is

$$Y_d(s) = y(0) + y(T_s)e^{-sT_s} + y(2T_s)e^{-2sT_s} + y(3T_s)e^{-3sT_s} + \cdots .$$

Introduce the notations $z = e^{sT_s}$ and $z^{-1} = e^{-sT_s}$, where $z = e^{sT_s}$ is the operator of the *z-transformation*. Then z^{-1} can be interpreted as a backward shift operator. Using the introduced notation, we have

$$Y_d(z) = y(0) + y(T_s)z^{-1} + y(2T_s)z^{-2} + y(3T_s)z^{-3} + \cdots = \sum_{k=0}^{\infty} y(kT_s)z^{-k}.$$

L. Keviczky et al., *Control Engineering: MATLAB Exercises*,
Advanced Textbooks in Control and Signal Processing,
https://doi.org/10.1007/978-981-10-8321-1_11

Example 11.1 Suppose given the *z-transform* of a signal as follows:

$$Y_{\mathrm{d}}(z) = \frac{2z^2 - z}{z^2 - z + 0.24}; \quad T_{\mathrm{s}} = 0.5.$$

Use MATLAB™ to determine the first 5 samples of the inverse *z-transform* of this signal.

(a) *Numerical calculation*

```
num=[2, -1, 0]
den=[1, -1, 0.24]
yd=dimpulse(num,den,5)
plot(yd,'*')
plot(1:5,yd,'*')
```

The data for this example can also be given in symbolic form using the *LTI sys* structure. Similarly to the way s was defined, a z variable can be defined by $z = e^{sT_{\mathrm{s}}}$. However, first the sampling time T_{s} should be specified:

```
Ts=0.5
z=tf('z',Ts)
Y=(2*z^2-z)/(z*z-z+0.24)
```

One advantage of using the *LTI sys* structure is that formally identical commands can be applied both for continuous-time and discrete-time systems. In the case of zero sampling time sys.Ts MATLAB™ considers the system as a continuous-time system, otherwise as a discrete-time system. The help command describes the exact usage of the impulse command:

```
help impulse
```

IMPULSE(SYS, TFINAL) simulates the impulse response of the system in the time range from $t = 0$ till the final time $t = \mathrm{TFINAL} = 2$. $\mathrm{TFINAL} = 2$ is the time required by the conditions of $T_s = 0.5$ and number of the samples (5). The impulse command essentially divides the numerator by the denominator and this is the way it provides the samples of the impulse response:

```
impulse(Y,2);
yd=impulse(Y,2);
plot(Ts*(0:4),yd,'*');
```

(b) *Partial fraction expansion*

Partial fractional form allows recognizing the components in the discrete time-domain. Then these components can simply be added up.

E.g. if an exponential component looks like

$$y(t) = e^{-at}; \quad Y_{\mathrm{d}}[nT_{\mathrm{s}}] = e^{-anT_{\mathrm{s}}} \xrightarrow{z} \frac{z}{z - e^{-aT_{\mathrm{s}}}},$$

then the partial fraction form contains components as follows:

$$Y_{\mathrm{d}}(z) = \frac{num}{den} = z\frac{num1}{den} = z\left(k_{\mathrm{o}} + \frac{r_1}{z - p_1} + \frac{r_2}{z - p_2}\right)$$

Then using the MATLAB™ command residue,

```
num1=[2, -1]
den=[1, -1, 0.24]
[r,p,k0]=residue(num1,den)
    r =
         1
         1
    p =
       0.6000
       0.4000
    k0 =
         []
Ts=0.5
```

$$Y_{\mathrm{d}}(z) = z\left(\frac{r_1}{z - p_1} + \frac{r_2}{z - p_2}\right) = z\left(\frac{1}{z - 0.6} + \frac{1}{z - 0.4}\right) = \frac{z}{z - 0.6} + \frac{z}{z - 0.4}$$

$$y_{\mathrm{d}}[k] = \{y(kT_{\mathrm{s}})\} = \left\{e^{akT_{\mathrm{s}}} + e^{bkT_{\mathrm{s}}}\right\}, \text{ where } e^{aT_{\mathrm{s}}} = 0.6; e^{bT_{\mathrm{s}}} = 0.4$$

```
a=log(0.6)/Ts
b=log(0.4)/Ts
```

(In MATLAB™ the command log means the calculation of the mathematical ln function.)

$$a = \frac{\ln(0.6)}{T_{\mathrm{s}}} = -1.02, \ b = \frac{\ln(0.4)}{T_{\mathrm{s}}} = -1.83$$

$$y(t) = e^{-1.02t} + e^{-1.83t}, \ t \geq 0.$$

```
t=[0:Ts:2]'
yd=exp(a*t)+exp(b*t)
plot(t,yd,'*');
```

11.1.2 Discrete-Time Impulse Response and Pulse Transfer Function

The discrete-time (pulse) transfer function describes the relation between the sampled output of a continuous-time process and the input of a holding unit driving the process (see Fig. 11.1). In practice, typically zero order holding (ZOH) units realized by analog-to-digital (A/D) converters are applied.

Example 11.2 Find the discrete-time pulse transfer function of the continuous process given by

$$P(s) = \frac{1}{(1+5s)(1+10s)}; \quad T_s = 2.5$$

supposing zero-order holding.

```
s=zpk('s')
Ps=1/((1+5*s)*(1+10*s))
```

Alternatively, the process can also be displayed in pole-zero form:

$$P(s) = \frac{0.02}{(s+0.2)(s+0.1)}$$

```
[zerof,polef,kf]=zpkdata(Ps,'v')
```

The process poles are: -0.2 and -0.1. Note that the process has no zeros.

MATLAB™ offers the c2d (read as continuous-to-discrete) command to derive the discrete-time pulse transfer function: $\mathsf{sysd} = \mathsf{c2d}(\mathsf{sysc}, \mathsf{ts}, \mathsf{method})$, where ts stands for the sampling time and method defines the type of the holding unit. The *default* method is 'zoh', i.e. zero-order holding, so it can be omitted.

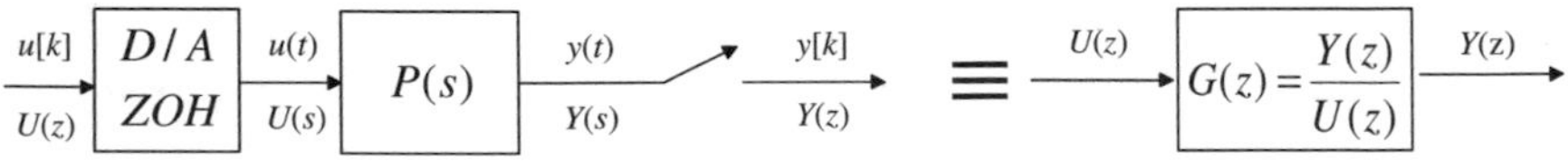

Fig. 11.1 Sampling and holding

```
Ts=2.5;
Gz=c2d(Ps,Ts,'zoh')
```

or alternatively:

```
Gz=c2d(Ps,Ts)
```

$$G(z) = 0.048929 \frac{(z+0.7788)}{(z-0.7788)(z-0.6065)}$$

Type

```
[zerod,poled,kd]= zpkdata(Gz,'v')
```

to get the discrete-time zeros and poles. The discrete-time poles are 0.7788 and 0.6065, while the discrete-time zero is −0.7788. Note that the discrete-time poles can be obtained from the continuous-time poles using the relation $z = e^{sT_s}$:

```
exp(-0.2*2.5)
     0.6065
```

Similarly

```
exp(polef*Ts)
```

can be used. Note that no similar simple relation exists between the continuous-time and discrete-time zeros.

Also note, that while the continuous-time process in the example has no zero, the discrete form still has one. This is because the sampling procedure results in subsidiary zeros. Processes of lag type exhibit several zeros, namely the number of discrete-time zeros will be less by one than the number of discrete-time poles.

11.1.3 Initial Value and Final Value Theorems

Once its z-transform is given, the initial value of a discrete-time signal can be calculated by:

$$\lim_{k \to 0} y(kT_s) = \lim_{z \to \infty} Y(z),$$

while the final value is obtained by

$$\lim_{k \to \infty} y(kT_s) = \lim_{z \to 1} \left(1 - z^{-1}\right) Y(z).$$

Apply a discrete-time unit step to drive the process of the previous example. The z-transform of the discrete-time unit step is $z/(z-1)$. Find the initial and final values of the sampled output.

$$Y(z) = \frac{z}{z-1} G(z)$$

The initial value of the sampled output can be obtained by $\lim_{z\to\infty} Y(z)$, which happens to be zero in this case (the degree of the denominator is higher than the degree of the numerator). The final value of the sampled output is obtained by substituting $z = 1$ into $G(z)$. This relation is in full harmony with finding the steady-state value of a step response produced by a continuous-time transfer function. MATLAB™ offers the dcgain command to find the dc gain both for continuous-time and discrete-time linear systems. For discrete-time systems ddcgain(numd, dend) can also be used. As far as the dcgain command is concerned, the sampling time will decide whether the system is of continuous-time or discrete-time.

```
Ps.Ts
Gz.Ts
A=dcgain(Ps)
Ad=dcgain(Gz)
```

As expected, both gains—A (CT) and Ad (DT)—will turn out to be 1.

11.1.4 Stability of Sampled-Data Systems

Discrete-time systems are stable if their poles (roots of the characteristic equation) are within the unit circle in the complex plane.

Example 11.3 Check the stability of the discrete-time system given by the following pulse transfer function:

$$G(z) = \frac{z^2 - 0.3z - 0.1}{z^3 + 3z^2 + 2.5z + 1}, T_s = 1$$

Define the discrete-time system for MATLAB™:

```
z=tf('z')
Gz=(z*z-0.3*z-0.1)/(z^3+3*z^2+2.5*z+1)
```

Consider the zeros and poles:

```
[zerod,poled,kd]=zpkdata(Gz,'v')
```

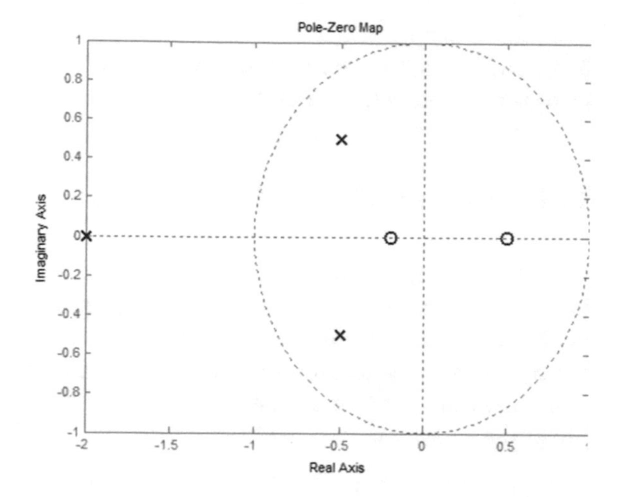

Fig. 11.2 Pole-zero configuration of the pulse transfer function

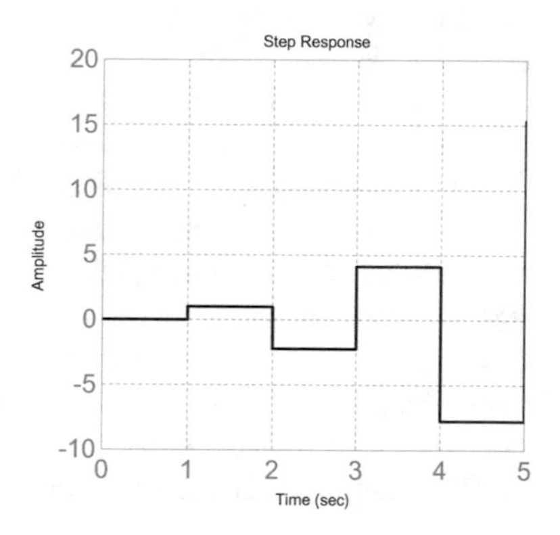

Fig. 11.3 The step response shows instability

The magnitude of the poles:

```
abs(poled)
   ans =
              2.0000
              0.7071
              0.7071
```

Since one pole is outside the unit circle, the system is unstable. Alternatively, this can be shown in graphical form as well (Fig. 11.2).

```
pzmap(Gz)
```

The zgrid command draws the unit circle together with the lines belonging to identical damping values and natural frequencies.

Stability can be checked in the time-domain by displaying the unit step response (see Fig. 11.3).

```
step(Gz),grid
```

Here the output is not bounded, so the system is unstable. Note that though MATLAB™ displays the samples together with zero-order holding, the impulse response of a discrete-time system consists of samples only.

11.2 Analysis of Closed-Loop Sampled-Data Systems

Closed-loop discrete-time systems may exhibit unexpected behaviour, as the system is essentially in open-loop between two samples. Also, the holding unit is an additional factor to modify the closed-loop behaviour. In the case of fast sampling, the continuous-time and discrete-time behaviours are not too far from each other. The following example illustrates the fundamental operation of a closed-loop sampled-data system.

Example 11.4 Assume that the continuous-time process to be controlled is an integrator given by the transfer function $1/s$. Unit negative feedback is applied (see Fig. 11.4) with sampling time T_s. A proportional controller with gain K provides the control signal. To convert the discrete-time control input train to a continuous-time signal, a zero-order holding unit is employed.

Analyze the dynamic behaviour of the closed-loop system and check the stability of the closed-loop system. Derive the difference equation of the closed-loop system and calculate the first 5 samples of the process output. Assume a unit step reference signal. Select $K = 1$ and repeat the analysis for various sampling times like $T_s = 0.5$, 1, 1.5 and 3. Calculate the impulse response transfer function for the closed-loop system and find an analytical form of the sampled output.

Figure 11.5 illustrates the nature of the error signal e and the output signal y.

The difference equation between the sampled output and the error signal is:

$$\begin{aligned} y[kT_s] &= y[(k-1)T_s] + KT_s e[(k-1)T_s] \\ &= y[(k-1)T_s] + KT_s\{r[(k-1)T_s] - y[(k-1)T_s]\} \end{aligned}$$

Here it has been taken into account that the error signal e is the difference between the reference signal r and the sampled output y. Rearranging this equation results in

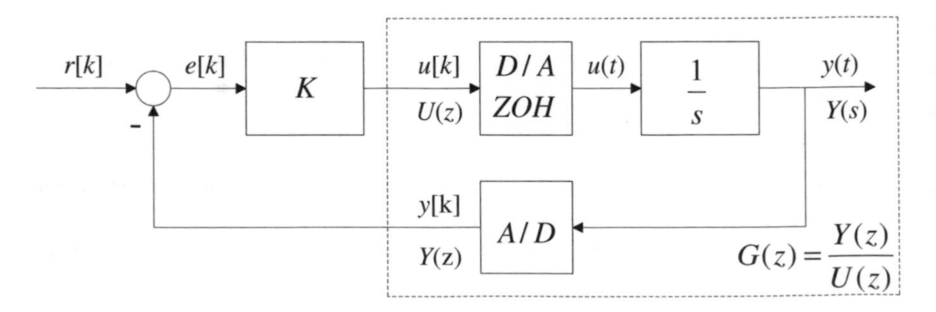

Fig. 11.4 Sampled control system

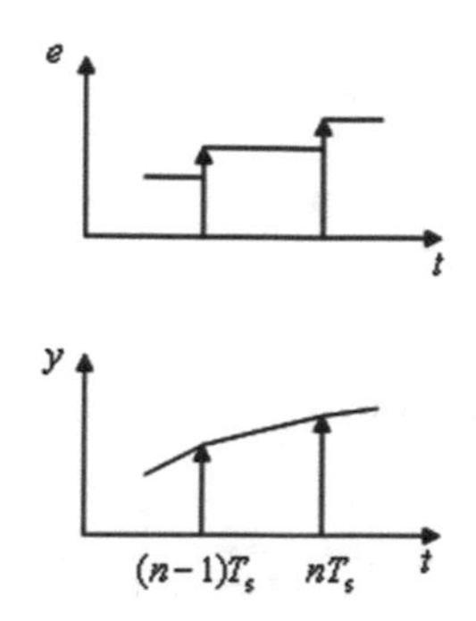

Fig. 11.5 Illustrating the error and the output signals

$$y[kT_s] = (1 - KT_s)y[(k-1)T_s] + KT_s r[(k-1)T_s]$$

or, further

$$\frac{y[kT_s] - y[(k-1)T_s]}{T_s} + Ky[(k-1)T_s] = Kr[(k-1)T_s],$$

which is the discretized version of the following differential equation as $T_s \to 0$:

$$\frac{\mathrm{d}y(t)}{\mathrm{d}t} + Ky(t) = Kr(t).$$

The continuous-time system is structurally stable, however, the discrete-time system exhibits various natures as the sampling time changes. The sequence of the output samples can be evaluated as follows:

For $T_s = 0.5$:	0; 0.5; 0.75; 0.875; 0.9375;	Stable, asymptotically converges to 1.
For $T_s = 1$:	0; 1; 1; 1; 1; 1;	Finite settling time.
For $T_s = 1.5$:	0; 1.5; 0.75; 1.125; 0.9375;	Stable, but oscillates.
For $T_s = 3$:	0; 3; −3; 9; −15;	Unstable.

Note that (unlike the continuous-time version of this problem) the discrete-time version will not be structurally stable.

The pulse transfer function of the closed-loop system is

$$T(z) = \frac{Y(z)}{R(z)} = \frac{\frac{KT_s}{z-1}}{1 + \frac{KT_s}{z-1}} = \frac{KT_s}{z - (1 - KT_s)}.$$

The poles of the closed-loop system remain within the unit circle if

$$|z_1| = |1 - KT_s| < 1.$$

The above condition can also be written as

$$0 < KT_s < 2$$

Introducing a new variable by $b = 1 - KT_s = e^{-aT_s}$, we have

$$Y(z) = \frac{z}{z-1}\frac{KT_s}{z-b}.$$

The partial fraction expansion is

$Y(z) = KT_s z\left(\frac{\alpha}{z-1} + \frac{\beta}{z-b}\right) = \frac{z}{z-1} - \frac{z}{z-b}$ where $\alpha = \frac{1}{1-b}$ and $\beta = \frac{1}{b-1}$.

Finally, the inverse z-transformation gives

$$y[kT_s] = 1[kT_s] - e^{-akT_s} = 1[kT_s] - \left(e^{-aT_s}\right)^k = 1[kT_s] - b^k.$$

Just to check this relation, assume $K = 1$ and $T_s = 0.5$:

$$b = 1 - K\,T_s = 0.5.$$

$$y[0] = 0; \quad y[T_s] = 1 - 0.5^1 = 0.5; \quad y[2T_s] = 1 - 0.5^2 = 0.75;$$
$$y[3T_s] = 1 - 0.5^3 = 0.875.$$

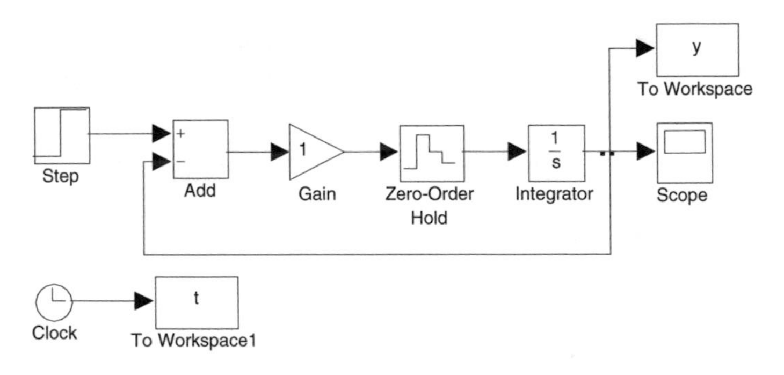

Fig. 11.6 SIMULINK™ diagram of the system

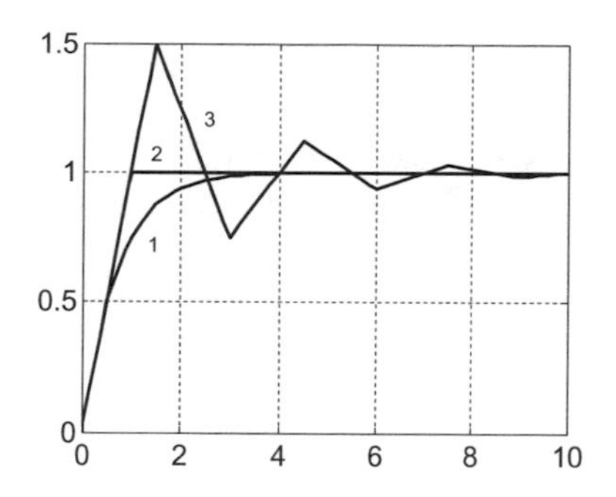

Fig. 11.7 The output signal for different sampling times

Alternatively, a SIMULINK™ program by Fig. 11.6 can be set up to illustrate the above calculations. Figure 11.7 shows the output for $T_s = 0.5$ (curve 1), $T_s = 1$ (curve 2) and $T_s = 1.5$ (curve 3).

11.3 State Space Equation of Sampled-Data Systems

The concept of the state-space model is valid for sampled-data systems, as well. The number of states describing the system dynamics is equal to the order of the difference equation representing the system. As learned earlier for continuous-time systems, several input-output equivalent state models can be derived for discrete-time systems, too.

11.3.1 Discretization of the Continuous-Time State Equation

Start with the solution of the continuous-time state-equation:

$$\boldsymbol{x}(t) = e^{\boldsymbol{A}(t-t_o)}\boldsymbol{x}(0) + \int_{t_o}^{t} e^{\boldsymbol{A}(t-\tau)}\boldsymbol{b}u(\tau)d\tau$$

Use a zero order holding unit. Enforce the integration between two consequtive sampling instants: $t_o = kT_s$ and $t = (k+1)T_s$. Within this region the input signal is constant: $u(\tau) = \text{constant} = u(kT_s)$. Assume that the $\boldsymbol{A}$ matrix is invertible. Then

$$\mathbf{x}[(k+1)T_s] = e^{AT_s}\boldsymbol{x}(kT_s) + \boldsymbol{A}^{-1}\left(e^{AT_s} - \boldsymbol{I}\right)\boldsymbol{b}u(kT_s).$$

Using the notation $u(kT_s) = u[k]$ and $\boldsymbol{x}[(k+1)T_s] = \boldsymbol{x}[k+1]$, the sampled version of the state equation takes the following form:

$$\boldsymbol{x}[k+1] = \boldsymbol{F}\boldsymbol{x}[k] + \boldsymbol{g}u[k]$$
$$y[k] = c^{\mathrm{T}}\boldsymbol{x}[k] + du[k]$$

where

$$\boldsymbol{F} = e^{AT_s} \quad \text{and} \quad \boldsymbol{g} = \boldsymbol{A}^{-1}\left(e^{AT_s} - \boldsymbol{I}\right)\boldsymbol{b}.$$

Note that there is no change in the parameters $\mathbf{c}^{\mathrm{T}}$ and d as the system is transformed to discrete form.

Example 11.5 Derive the state equation for the system introduced in Example 11.2. Define the transfer function of the continuous-time process, and for sampling employ sampling time $T_s = 2.5$.

$$P(s) = \frac{1}{(1+5s)(1+10s)}$$

The related MATLAB™ commands are:

```
s=zpk('s')
Ps=1/((1+5*s)*(1+10*s))
Sysc=ss(Ps)
[A,b,c,d]=ssdata(Sysc)
Ts=2.5
Sysd=c2d(Sysc,Ts,'zoh')
[F,g,c1,d1]=ssdata(Sysd)
%check
F1=expm(A*Ts)
g1=inv(A)*(expm(A*Ts)-eye(2))*b
c2=c
d2=d
% display the continuous-time and
% the discrete-time step responses
step(Sysc,Sysd)
```

Summing up the results, the parameter matrices of the continuous-time system are:

```
A = -0.1000   1.0000
    0        -0.2000
b =  0
     0.1250
c =  0.1600   0
d =  0
```

while the parameter matrices of the discretized system are:

```
F =  0.7788   1.7227
     0        0.6065
g =  0.3058
     0.2459
c1 = 0.1600    0
d1 = 0
```

Application of the analytical relationships leads to the same result.

11.3.2 Derivation of the Discrete State Equation from the Pulse Transfer Function

Several state representations can be derived from the pulse transfer function.

Example 11.6 Consider the system introduced in Example 11.5. The related MATLAB™ program is:

```
% starting from a zero-pole-gain form
s=zpk('s')
Ps=1/((1+5*s)*(1+10*s))
Ts=2.5
z=zpk('z',Ts)
Pz=c2d(Ps,Ts)
Sysz=ss(Pz)
[F,g,c,d]=ssdata(Sysz)
% starting from a polinom/polinom form
s=tf('s')
Ps=1/((1+5*s)*(1+10*s))
Ts=2.5
Pz1=c2d(Ps,Ts)
Sysz1=ss(Pz1)
[F1,g1,c1,d1]=ssdata(Sysz1)
% canonical form
Sysz2=canon(Sysz1,'modal')
```

```
[F2,g2,c2,d2]=ssdata(Sysz2)
figure(1)
step(Sysz,Sysz1,Sysz2)
```

Summarizing the results:

```
        0.048929 (z+0.7788)
Pz= ---------------------
     (z-0.7788) (z-0.6065)


F =  0.6065   1.1770
     0          0.7788
g =  0
     0.2500
c =   0.2304    0.1957
d =   0

       0.04893 z + 0.03811
Pz1= ----------------------
       z^2 - 1.385 z + 0.4724


F1 = 1.3853  -0.4724
      1.0000      0
g1 = 0.2500
      0
c1 = 0.1957   0.1524
d1 = 0
Parameter matrices of the canonical form:
F2 = 0.7788   0
      0      0.6065
g2 = 2.6862
    -2.2815
c2 = 0.1647   0.1725
d2 = 0
```

The parameter matrices belonging to different interpretations are not identical, though the pulse transfer function behind these interpretations is unique. The second interpretation is the so called controllability form.

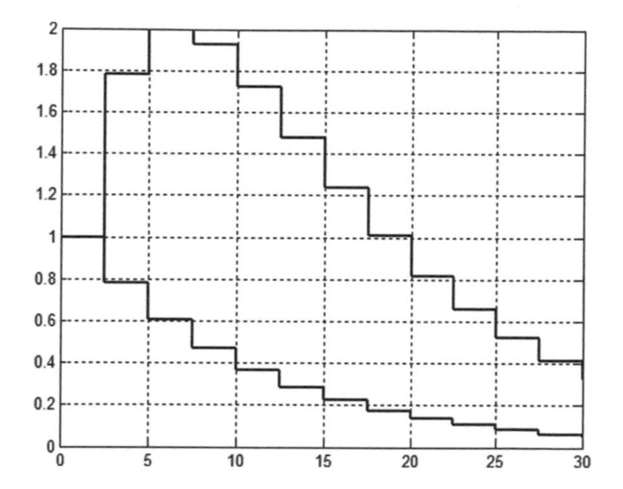

Fig. 11.8 The time course of the state variables in the sampled system

Investigate the system response to various initial conditions. Set $\boldsymbol{x}[0] = [1 \quad 1]^{\mathrm{T}}$ and consider the first state-space representation.

The MATLAB™ program is

```
s=zpk('s')
Ps=1/((1+5*s)*(1+10*s))
Ts=2.5
z=zpk('z',Ts)
Pz=c2d(Ps,Ts)
Sysz=ss(Pz)
[F,g,c,d]=ssdata(Sysz)
x0=[1;1]
tfinal=30;
figure(2)
[y,t,x]=initial(Sysz,x0,tfinal)
stairs(t,x),grid
```

The state variables versus the time are shown in Fig. 11.8.

Chapter 12
Discrete Regulator Design for Stable Processes

12.1 Design of a Youla Parameterized Regulator

The Youla parameterized control can be applied also to sampled systems. The regulator design is similar to the procedure discussed in Chap. 7. In sampled data systems, the realization of the regulator for processes with dead-time does not cause any problems.

The sampled system is given by its pulse transfer function G. The pulse transfer function of the process has to be separated into the cancellable G_+ and the non-cancellable G_- components. The discrete dead-time is denoted by d.

$$G = G_+ G_- z^{-d}$$

In the case of lag elements there is always a z^{-1} term, so the discrete dead-time d is calculated as the ratio of the real, physical dead-time T_d and the sampling time T_s plus one, i.e., $d = \mathrm{entier}(T_\mathrm{d}/T_\mathrm{s}) + 1$. Preferably choose the sampling time so that the ratio $T_\mathrm{d}/T_\mathrm{s}$ is an integer.

R_r is the pulse transfer function of the reference model (reference filter) and R_n is the pulse transfer function of the disturbance filter.

The expression of the Youla parameter is: $Q = R_\mathrm{n} G_+^{-1}$.

The block diagram of the Youla parameterized discrete control system in *IMC* form is given in Fig. 12.1. (An equivalent circuit is shown in Fig. 12.3 of the textbook [1].)

If the disturbance is zero and the process and its model are the same, then the feedback signal is zero, and the reference signal tracking is realized according to

$$y = R_\mathrm{r} G_- z^{-d} r.$$

L. Keviczky et al., *Control Engineering: MATLAB Exercises*,
Advanced Textbooks in Control and Signal Processing,
https://doi.org/10.1007/978-981-10-8321-1_12

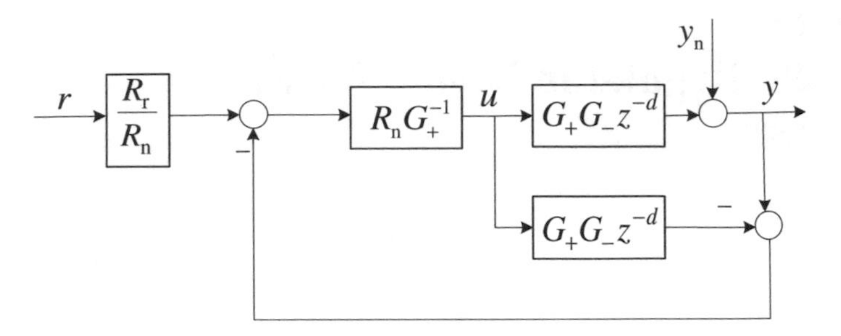

Fig. 12.1 Block diagram of the YOULA parameterized discrete control system

For reference signal tracking without static error, the static gain of G_- and R_r should be 1, i.e. $G_-(z=1)=1$ and $R_r(z=1)=1$. The resulting transfer function for the disturbance input is

$$y = \left(1 - R_n G_- z^{-d}\right) y_n.$$

To ensure a zero steady output value in case of a step disturbance (total disturbance rejection) the static gain of R_n should be also 1, i.e. $R_n(z=1)=1$.

The zero of the process which is outside of the unit circle should not be cancelled, as this would result in an unstable pole in the regulator. Also it is not expedient to cancel zeros which are on the left side of the unit circle, as this would introduce a pole in the regulator which would cause intersampling oscillations in the control system. (This phenomenon will be analysed in more detail in the discussion of dead-beat control.)

Example 12.1 Consider the dead-time process analysed in Example 7.3.

The transfer function of the process is

$$P(s) = \frac{1}{(1+5s)(1+10s)} e^{-30s}.$$

The sampling time is $T_s = 1$ s.

Design a YOULA parameterized regulator for this process. First the filters are chosen to provide a delay of one sampling time. Then analyse how is the behaviour of the control system modified if the filters are obtained by sampling the continuous filters given by the transfer function $R_r(s) = R_n(s) = \frac{1}{(1+s)^2}$.

Write a MATLAB™ program whose input data are the invertible and non-invertible factors G_+ and G_- of the pulse transfer function of the process, and the filters are R_r and R_n. The program has to calculate and plot the output and control signals of the control system for unit step reference signal and zero output disturbance, then the output and the control signals for zero reference signal and unit step output disturbance.

Save the program with the name Youla_discrete.

```
% Youla_discrete: Youla discrete basic program
display('.....Q='),Q=minreal(Rn/Gp,0.0001)
display('.....C='),C=minreal(Q/(1-Q*G),0.0001)
display('.....L='),L=minreal(C*G,0.0001)
display('.....Tr='),Tr=minreal((Rr/Rn)*Q*G,0.0001)
display('.....Ur='),Ur=minreal((Rr/Rn)*Q,0.0001)
pause
t=0:Ts:50;
figure(1)
yr=step(Tr,t);
subplot(211), plot(t,yr,'*'),grid
ur=step(Ur,t);
subplot(212), stairs(t,ur),grid
pause;
display('.....Sn='),Sn=minreal((1-Q*G),0.0001)
display('.....Un='),Un=minreal(-C*(1-Q*G),0.0001)
pause
figure(2)
yn=step(Sn,t);
subplot(211),plot(t,yn,'*'),grid;
un=step(Un,t);
subplot(212),stairs(t,un),grid;
```

Give the process in MATLAB™.

```
clear; clc; s=zpk('s')
P=1/((1+5*s)*(1+10*s))
Ts=1; z=zpk('z',Ts); G1=c2d(P,Ts)
G=G1*z^(-30)
```

The pulse transfer function $G(z)$ considering also the dead-time is

$$\frac{0.0090559\,(z+0.9048)}{(z-0.9048)\,(z-0.8187)}\,z^{-30}$$

Separate the pulse transfer function of the process into cancellable and non-cancellable factors. First suppose that the whole dynamics can be cancelled.

```
Gm=1;          %G_
Gp=G1/Gm       %G+
```

Set the filters as

```
Rr=1/z; Rn=1/z;
```

Call the Youla_discrete program.

```
Youla_discrete
```

Figure 12.2 gives the output signal (upper figure) and the control signal (lower figure) for unit step reference signal. Figure 12.3 shows the output and the control signals for unit step output disturbance. It is a strange phenomenon that the oscillating control signal (on the lower figures) results in a calm output signal. The explanation is that the simulation is executed only at the sampling points. Simulating the real continuous process decreasing oscillations can be observed between the sampling points.

Build the SIMULINK™ block diagram corresponding to Fig. 12.1. Let the process be considered with its continuous model (Fig. 12.4). Running the simulation it can be seen that really there are oscillations between the sampling points (Fig. 12.5).

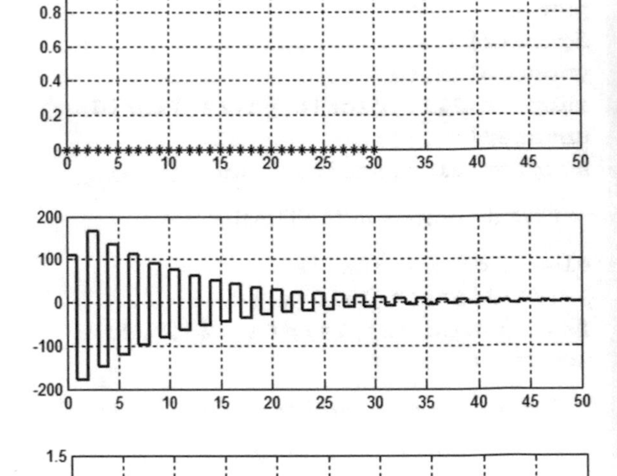

Fig. 12.2 Output and control signals for step reference signal

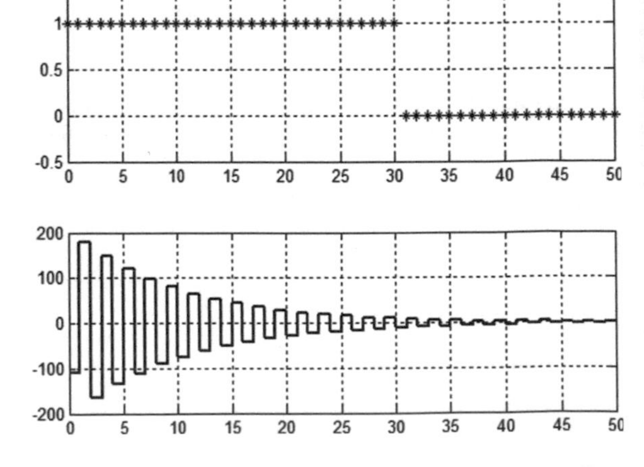

Fig. 12.3 Output and control signals for step output disturbance signal

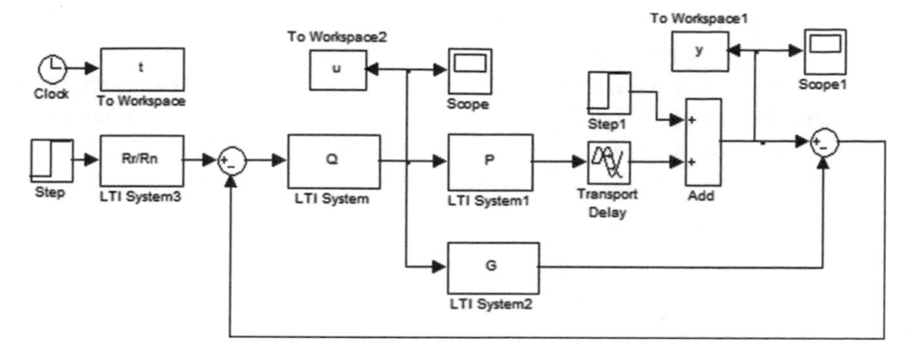

Fig. 12.4 SIMULINK™ diagram for the Youla parameterized control system

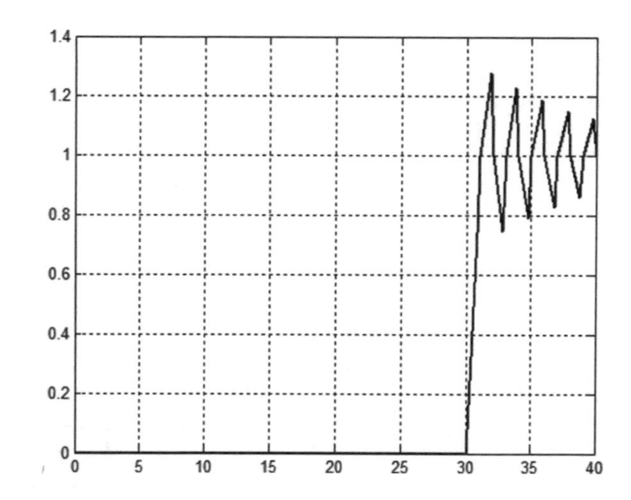

Fig. 12.5 Oscillations between the sampling points

In the following the non-cancellable part contains a zero on the left side of the unit circle.

The separation is as follows:

$$G_{-} = \frac{1+0.9048z^{-1}}{1.9048}; \quad G_{+} = \frac{0.0090559 \cdot 1.9048}{(1-0.9048z^{-1})(1-0.8187z^{-1})}.$$

With MATLAB commands:

```
Gm =(1+0.9048*z^(-1))/1.9048
Gp = minreal(G1/Gm,0.0001)
```

Then call the program:

```
Youla_discrete
```

Figure 12.6 shows the output and the control signal for unit step reference signal. The signals for a disturbance input are shown in Fig. 12.7. The control signal reaches a steady value after two jumps, and the output signal shows calm behaviour. Running the SIMULINK™ model it can be seen that at the output of the continuous process there are no oscillations between the sampling points.

Investigate the operation of the control system with second order filters.

```
Rr=c2d(1/(1+s)^2,Ts);Rn=Rr;
```

Then run the program again.

```
Youla_discrete
```

In Figs. 12.8 and 12.9 it can be seen that the dynamics of the settling process has changed, it became a bit slower.

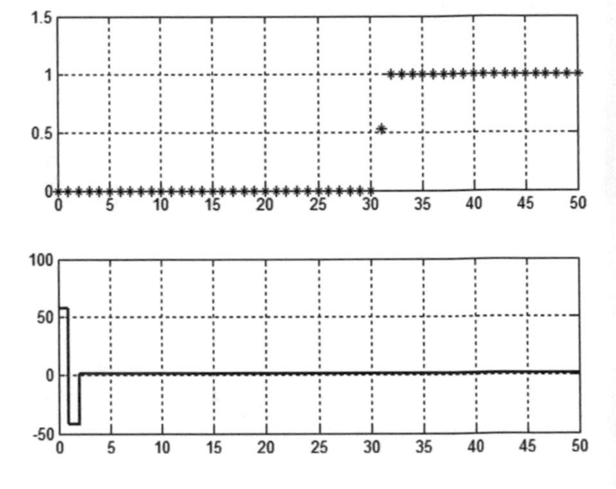

Fig. 12.6 Output and control signals for step reference signal with appropriate separation of G

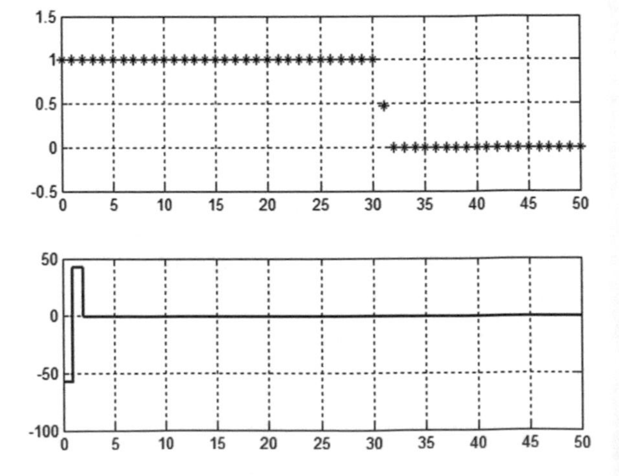

Fig. 12.7 Output and control signals for step output disturbance with appropriate separation of G

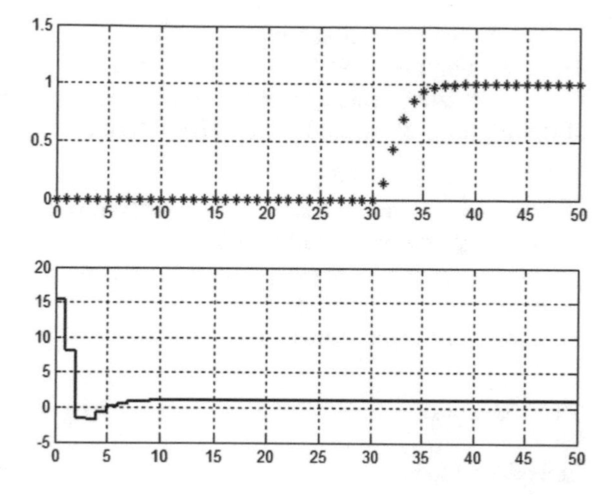

Fig. 12.8 Output and control signals for step reference signal with filters

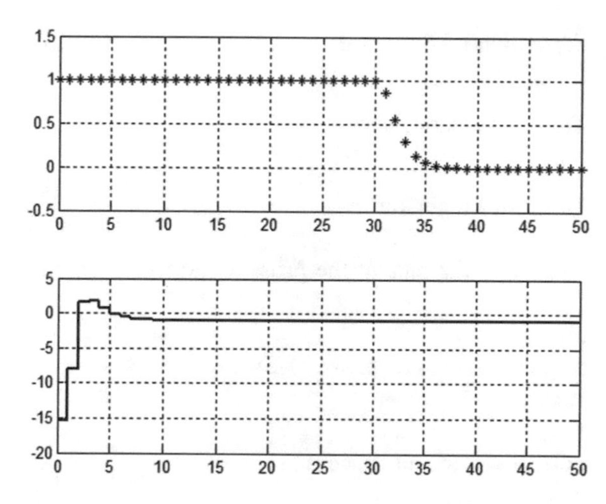

Fig. 12.9 Output and control signals for step output disturbance with filters

The role of filters on the one hand is to influence the dynamic behaviour of the control system. If the two filters are different, the dynamics of reference signal tracking and that of disturbance rejection will be different. In this case it is called *two-degree-of-freedom* (*2DOF*) control. On the other hand, applying the filters, the maximum value of the control signal becomes smaller, as can be seen in Figs. 12.6, 12.7, 12.8 and 12.9. The filters also influence the robustness of the control system. If the process and its model are not exactly the same, i.e. there is a plant-model mismatch, by choosing the appropriate filters generally robust behaviour of the control system can be achieved, i.e. the behaviour of the control system can be acceptable even if the model is not accurate.

Problem Analyse the reference signal tracking and disturbance rejection properties of the control system if the dynamics of the two filters differs.

Execute the simulation with the SIMULINK™ model, both for the case where the process and its model are the same and also if there is a mismatch between them. In the latter case let the dead-time of the system be 40 s, while the dead-time of the model is 30 s. Find filters which ensure acceptable behaviour even in this case.

Example 12.2 Consider Example 12.1 in the text book [1]. Simulate the behaviour of the control system in MATLAB™.

The process is a sampled first order lag element with dead-time.

The pulse transfer function of the process is: $G(z) = \dfrac{0.2}{z-0.8} z^{-3} = G_+ G_- z^{-d}$.

The pulse transfer functions of the filters are: $R_r(z) = \dfrac{0.8z^{-1}}{1-0.2z^{-1}}$ and $R_n(z) = \dfrac{0.5z^{-1}}{1-0.5z^{-1}}$.

Embedded filters: $G_r = G_n = 1$.

Separation of the process: $G_+(z) = \dfrac{0.2z^{-1}}{1-0.8z^{-1}}$ and $G_-(z) = 1$.

The YOULA parameter: $Q = R_n G_+^{-1} = \dfrac{2.5(1-0.8z^{-1})}{1-0.5z^{-1}}$.

The series regulator: $C = \dfrac{R_n G_+^{-1}}{1 - R_n G_- z^{-3}} = \dfrac{2.5(1-0.8z^{-1})}{1-0.5z^{-1}-0.5z^{-4}}$.

Specify the data of the process and the filters in MATLAB™.

```
z=zpk('z');
Gp=0.2*z^(-1)/(1-0.8*z^(-1));          % G+
Gm=1;                                  % G_
d=3;
G1=minreal(Gp*Gm,0.0001);
G=G1*z^(-d);
Rr=0.8*z^(-1)/(1-0.2*z^(-1));
Rn=0.5*z^(-1)/(1-0.5*z^(-1));
```

Give a fictive sampling time: **Ts=1;**
Then call the program Youla_discrete:

```
Youla_discrete
```

Figure 12.10 shows the dynamics of the reference signal tracking, while Fig. 12.11 gives the dynamics of the disturbance rejection.

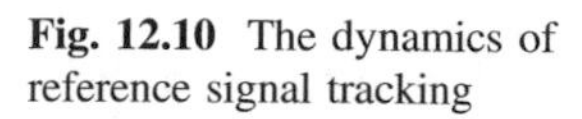

Fig. 12.10 The dynamics of reference signal tracking

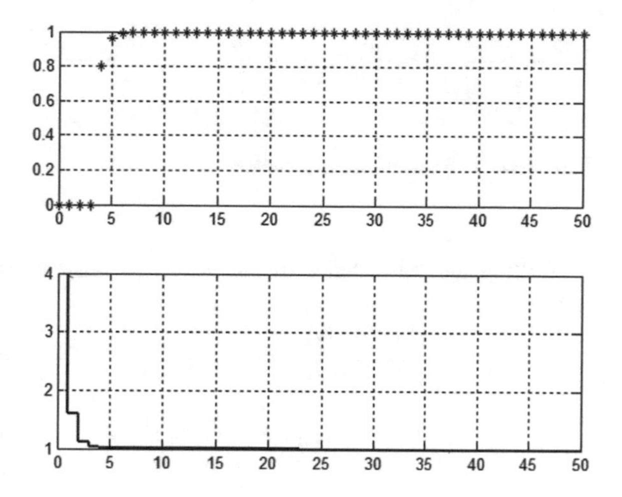

Fig. 12.11 The dynamics of disturbance rejection

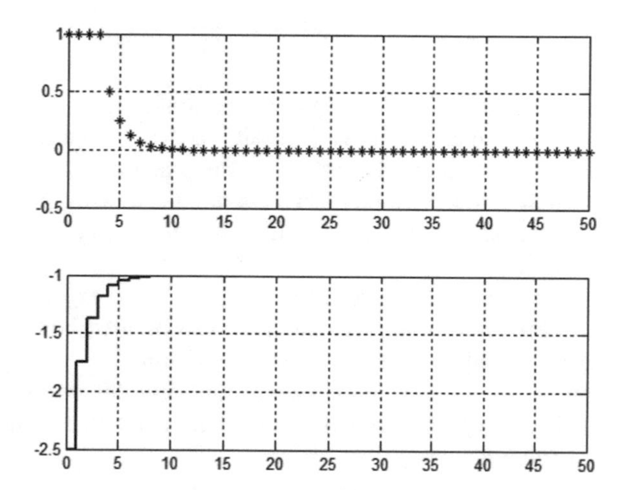

12.2 Control of a Dead-Time System with a SMITH Predictor

If the process contains a large dead-time, applying a *PID* regulator, the control system will be slow (this behaviour was demonstrated in the design of a continuous regulator in Chap. 8). To accelerate the control system, a YOULA parameterized regulator can be used, or the SMITH predictor regulator developed earlier, around 1950. The design of a SMITH predictor brings up the question, whether the usual control system of a process with dead-time could be made equivalent to a control system where the dead-time appears outside of the closed loop (Fig. 12.12).

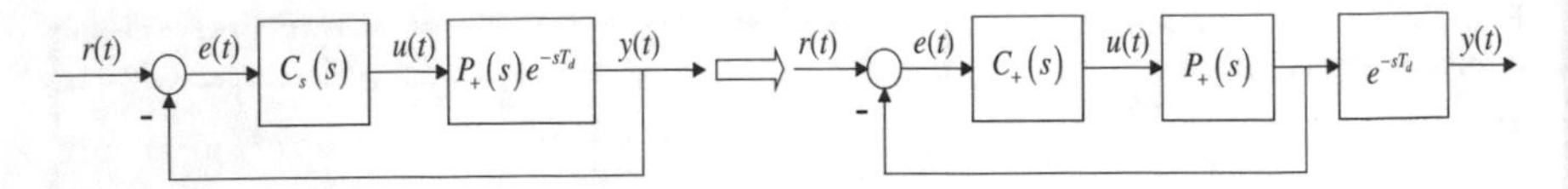

Fig. 12.12 The idea of Smith predictor

Writing the equivalence of the resulting transfer functions of the two systems yields

$$\frac{C_s(s)P_+(s)e^{-sT_d}}{1+C_s(s)P_+(s)e^{-sT_d}} = \frac{C_+(s)P_+(s)}{1+C_+(s)P_+(s)}e^{-sT_d}$$

$$C_s(s)[1+C_+(s)P_+(s)] = C_+(s)[1+C_s(s)P_+(s)e^{-sT_d}]$$

whence

$$C_s(s) = \frac{C_+(s)}{1+(1-e^{-sT_d})C_+(s)P_+(s)}$$

is the transfer function of the Smith predictor.

The regulator $C_+(s)$ is designed for the process without dead-time. This control system will be fast. Then the practically applicable regulator is calculated according to the relation above, giving the transfer function of the Smith predictor. (Let us remark that the signal of the process output before the dead-time is not available.)

It is seen that the dead-time appears in the expression of the regulator.

Theoretically, a Smith predictor can be used both for continuous and sampled systems, but as the realization of dead-time in continuous systems is difficult and can be solved only approximately, therefore this algorithm generally is applied only in discrete systems. In sampled systems, dead-time means the shift of the signal and therefore it is simply realizable.

The pulse transfer function of the discrete Smith predictor is

$$C_s(z) = \frac{C_+(z)}{1+(1-z^{-d})C_+(z)G_+(z)}$$

where $G_+(z)$ is the pulse transfer function of the process without the dead-time, and d is the discrete dead-time.

Example 12.3 Design a Smith predictor regulator for a proportional process containing three time lags and dead-time. The transfer function of the continuous process is

$$P(s) = P_+(s)\,e^{-sT_d} = \frac{K\,e^{-sT_d}}{(1+sT_1)\,(1+sT_2)\,(1+sT_3)},$$

with parameters $K=5$, $T_1=10$, $T_2=4$, $T_3=1$ and $T_d=10$.

Design the regulator $C_+(s)$ to ensure a phase margin of about 60°. The maximum value of the control signal should not exceed the value umax = 10.

```
K=5; T1=10; T2=4; T3=1; Td=10; ft=60;
```

Define the variable 's' by

```
s=zpk('s');
```

Use the *LTI* structures of the *Control System Toolbox* ($P_+(s)$ = Ps)

```
Ps=1/((1+s*T1)*(1+s*T2)*(1+s*T3))
```

First Step: Design of the $C_+(z)$ Discrete Regulator

A good practical rule for the choice of the sampling time is that it should be smaller than the smallest time constant of the process. (For simulation tasks it could be chosen to be about one tenth of the smallest time constant, for control purposes it is appropriate if the sampling time is chosen around one half or one third of the smallest time constant; at most it can be equal to it. Besides, it is expedient to choose the sampling time in such a way that the ratio of the dead-time and the sampling time is an integer.)

```
Ts=0.5;
```

The regulator $C_+(z)$ is designed in the frequency domain according to the low frequency approximation method described in Chap. 13.

The pulse transfer function of the sampled continuous process with a zero-order holding ($G_+(z)$ = Gz) is

```
Gz=c2d(Ps,Ts)
```

The resulting pulse transfer function is

$$G_+(z) = 0.0004416 \frac{(z+0.2254)(z+3.167)}{(z-0.9512)(z-0.8825)(z-0.6065)}.$$

The *PID* regulator based on pole cancellation is designed according to Chap. 13.

The pole of $G_+(z)$ belonging to the biggest time constant, $p_1 = e^{-T_s/T_1} = 0.9512$ is cancelled and instead an integrator is introduced (*PI*), and the pole belonging to the second biggest time constant $p_2 = -0.8825$ is also cancelled, and instead an ideal differentiating effect is introduced (*PD*). So the pulse transfer function of the regulator is

$$C_{+}(z) = k_c \frac{z - 0.9512}{z - 1} \frac{z - 0.8825}{z}.$$

The constant k_c in the regulator is designed to set the prescribed phase margin. In the first step, let its value be 1.

```
p1=exp(-Ts/T1)
p2=exp(-Ts/T2)
kc=1;
Cz=zpk([p1 p2],[0 1],kc,Ts)
```

The pulse transfer function of the open loop is

```
Lz=Cz*Gz
```

Executing the simplifications:

```
Lz=minreal(Lz)
```

The value of k_c can be read from the graphical surface of *ltiview* viewer. Its value will be the reciprocal of the gain belonging to the phase angle $-120°$.

```
ltiview('bode',Lz);
```

or k_c can be calculated directly with command margin,

```
[mag,phase,w]=bode(Lz);
kc=margin(mag,phase-60,w);
```

The gain is $k_c = 33.46$.
Calculate again the transfer functions:

```
Cz=kc*Cz;
Lz=kc*Lz;
```

Check the phase margin:

```
margin(Lz)
```

Calculate and plot the output and control signals of the discrete system.

```
Hz=Lz/(1+Lz)
Uz=Cz/(1+Lz)
subplot(211); step(Hz)
subplot(212); step(Uz)
```

Second Step: Calculation of the Pulse Transfer Function $C_{sm}(z)$ of the SMITH Predictor

$$C_{sm}(z) = \frac{C_+(z)}{1 + (1 - z^{-d})C_+(z)G_+(z)}$$

where $d = T_d/T_s$

```
z=tf('z',Ts);
d=Td/Ts
Csm=Cz/(1+(1-z^(-d))*Cz*Gz)
Csm=minreal(Csm,0.0001)
```

The behaviour of a hybrid (discrete–continuous) system can be analysed in the SIMULINK™ environment. Build the SIMULINK™ model (Fig. 12.13.). The model takes the parameters Csm, Ps from the MATLAB™ surface. The sampling time is set at the zero order hold element. The hold element can be left out of the circuit, as when connecting the discrete and continuous blocks SIMULINK™ automatically employes a zero order hold. Set the value of the dead-time. Execute the simulation till 30 s. The result of the simulation can be seen on the scopes, but the simulation data can be reached also in MATLAB™. Double clicking on the *Scope* blocks, choose the option *Parameters* (the second icon from the left). Then on the option *Data History* mark *Save data to workspace*, then set the *Variable name* to ty and tu, respectively on the scopes with format *Array*. After running the SIMULINK™ model, the signals can be plotted from MATLAB™ with the following commands:

```
subplot(211);plot(ty(:,1),ty(:,2));grid;
subplot(212);stairs(tu(:,1),tu(:,2));grid;
```

Determine the overshoot and the settling time of the output signal and check the maximum value of the control signal. The output and the control signals are shown in Fig. 12.14. The control is much faster than it would be with a series *PID* regulator designed for the process with dead-time.

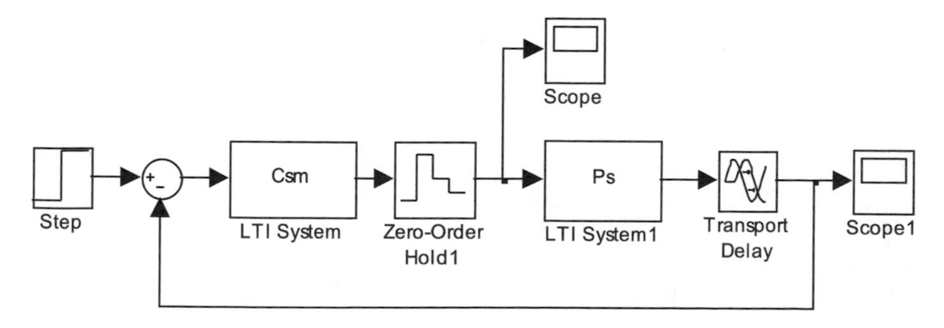

Fig. 12.13 SIMULINK™ diagram of SMITH predictor

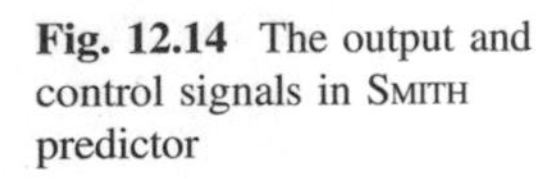

Fig. 12.14 The output and control signals in SMITH predictor

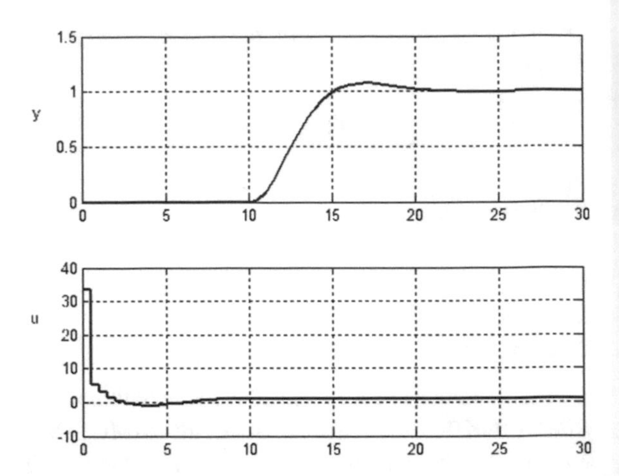

Observe that the control signal is the same as the control signal obtained in the first step of the design for the case of the process without dead-time. The output signal is the same as the output signal of the control without the dead-time, but shifted by the dead-time.

Problem Design a SMITH predictor regulator for the second order process with dead-time given in Example 12.1. For the process without dead-time design a *PID* regulator with phase margin of 60°.

Remark The SMITH prediktor is a special case of the YOULA parameterization. It does not apply filters.

Problem Determine the expression of the YOULA parameter corresponding to the SMITH predictor.

Problem With the SIMULINK™ model execute simulations for the case of a plant-model mismatch. Change the parameters of the continuous process (static gain, time constants, dead-time one by one and also together) by ±10%. The regulator was designed for the original, nominal process. Evaluate the simulation results.

12.3 Design of a Dead-Beat Regulator

The structure of the control system is shown in Fig. 12.15.

In sampled-data (discrete) control systems, the regulator is an intelligent device, typically it is a microprocessor based *Programmable Logic Regulator* (*PLC*) device. The task of the *PLC* is the realization of a control algorithm, and handling of signals related to the operation of the regulator (filtering, A/D and D/A converters, interfaces). With software realization, besides *PID* control there is the possibility of applying several special control algorithms. Such algorithms are the

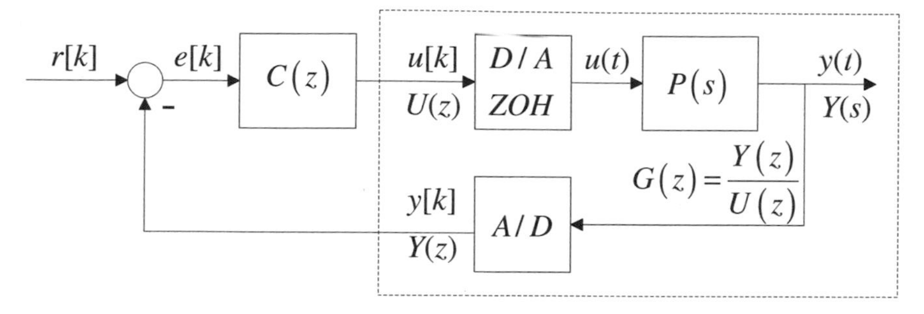

Fig. 12.15 Block diagram of a sampled control system

discrete *PID* regulator, the YOULA parameterized regulator and the SMITH predictor. Another discrete control algorithm is the dead beat regulator, which ensures the accurate settling of the output signal within a given finite number of the sampling periods. First analyse the method applying it to stable processes without dead-time ($T_d = 0$), and the reference signal is supposed to be a unit step. Let us remark that dead beat regulator can be designed also for cases when these conditions are not fulfilled.

In the sequel the solution is shown in three steps. In the first step the fastest control is designed, the requirement is to settle the process output to the desired value in one sampling step. It will be seen that settling in one step can be ensured, but the control signal values can be extremely high, furthermore, in most cases intersampling oscillations do appear. In the second step, the cancellable and non-cancellable zeros will be separated to avoid oscillations in the pulse transfer of the process. It will be seen that the oscillations can be avoided by increasing the prescribed settling time. If after that the design process still results in control signals that are too high, then in the third step, the design procedure can be refined by introducing a design polynomial, increasing the settling time, but still keeping it finite.

The design is executed in the z domain. An interesting feature of the method is that it eliminated some non-desired time domain properties of the system (oscillations, high overexcitation) taking into consideration properties in the z domain. The basic task is the design of the discrete regulator.

First the hybrid (continuous-discrete) problem is converted to a pure discrete problem by determining the pulse transfer function $G(z)$, which is the discrete equivalent of the connected D/A converter and holding element and continuous process given by its transfer function $P(s)$. The sampling time T_s also has to be given. Then the pulse transfer function $C(z)$ of the regulator is designed and the behaviour of the closed loop control system is analysed.

The main point of the design is that we prescribe the behaviour of the closed loop giving its closed loop pulse transfer function $T(z)$. In the case of dead beat control the resulting transfer function ensures the settling after a unit step reference signal during a given number of sampling steps, i.e. $T(z) = z^{-d}$, where d is the discrete dead-time, namely $d = \text{entier}(T_d/T_s) + 1$.

$$\frac{C(z)\,G(z)}{1 + C(z)\,G(z)} = T(z)$$

Hence $C(z)$, the pulse transfer function of the regulator, is

$$C(z) = \frac{T(z)}{G(z)\,[1 - T(z)]}.$$

Example 12.4 Consider the continuous process given by the transfer function

$$P(s) = \frac{1}{(1+5s)\,(1+10s)}$$

The sampling time is $T_s = 1$ s.

First define the s and z variables.

```
Ts=1;
s=zpk('s');
z= zpk ('z',Ts);
Ps=1/((1+5*s)*(1+10*s))
```

To have an idea what the requirement of settling in one sampling period means draw the step response of the process.

```
step(Ps);
```

Calculate the pulse transfer function of the process:

```
Gz=c2d(Ps,Ts)
```

The obtained pulse transfer function is

$$G(z) = \frac{\mathcal{B}(z)}{\mathcal{A}(z)} = 0.0090559\frac{(z+0.9048)}{(z-0.8187)\,(z-0.9048)}$$

The discrete poles are transformed from the continuous poles according to the relationship $z_i = e^{p_i T_s} = e^{-T_s/T_i}$. A discrete zero also appeared.

First design a one step dead-beat regulator. The resulting transfer function between the output signal and the reference signal should be

$$T(z) = z^{-1}.$$

The regulator is obtained as

$$C(z) = \frac{1}{G(z)\,(z-1)}.$$

```
Tz=1/z
Cz=Tz/(Gz*(1-Tz))
Cz=minreal(Cz)
```

The pulse transfer function of the regulator is

$$C(z) = 110.425\,\frac{(z-0.8187)\,(z-0.9048)}{(z+0.9048)\,(z-1)}.$$

Analyse the behaviour of the control system in discrete time. The step response really shows a one step delay.

```
step(Cz*Gz/(1+Cz*Gz))
```

If we want to get an accurate picture of the behaviour of the system, the behaviour of the continuous process should be analysed considering also the output signal values between the sampling points. The simulation is not so straightforward in the MATLAB™ environment, but it is easy with SIMULINK™. Start SIMULINK™ and build the model shown in Fig. 12.16.

```
simulink
```

Create a new model file (with extension *.mdl*) and copy the individual blocks from the block libraries. Change the values of the parameters to the required values. From the menu set the *Simulation –> Parameters –> Stop* time parameter to 8. SIMULINK™ uses the parameters defined in MATLAB™.

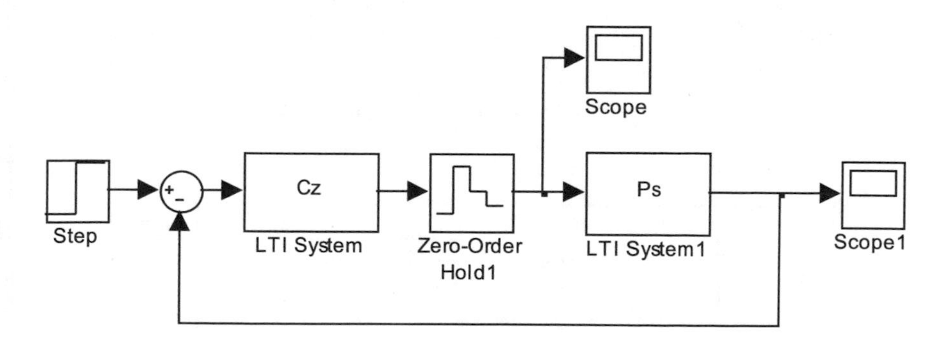

Fig. 12.16 SIMULINK™ diagram of a sampled control system

The individual blocks are taken from the following block libraries:

$C(z)$, $P(s)$: *Control System Toolbox -> LTI system*
Zero order hold: *Simulink -> Discrete -> Zero-Order-Hold*
Subtraction: *Simulink -> Math -> Sum*
Step input: *Simulink -> Sources -> Step*
Scope: *Simulink -> Sinks -> Scope*.

The zero order hold element can be left out of the block diagram. The results can be seen on the *Scope* blocks.

The results can be transformed to MATLAB™ by double clicking on the *Scope* blocks, choosing the option *Parameters*, and then indicating on *Data History* the option *Save data to workspace.* Set *Variable name* ty and tu, respectively, with *Array* format.

After running the SIMULINK™ model, the signals can be plotted from the MATLAB™ environment by the following commands

```
subplot(211);plot(ty(:,1), ty(:,2));grid;
subplot(212);stairs(tu(:,1), tu(:,2));grid;
```

The simulation result shown in Fig. 12.17 demonstrates that there are oscillations between the sampling points. The reason for the oscillations is that the regulator compensated the numerator $\mathcal{B}(z)$ of the process, and therefore it has a pole which causes oscillations in the signal $u[k]$. If the pulse transfer of the process is

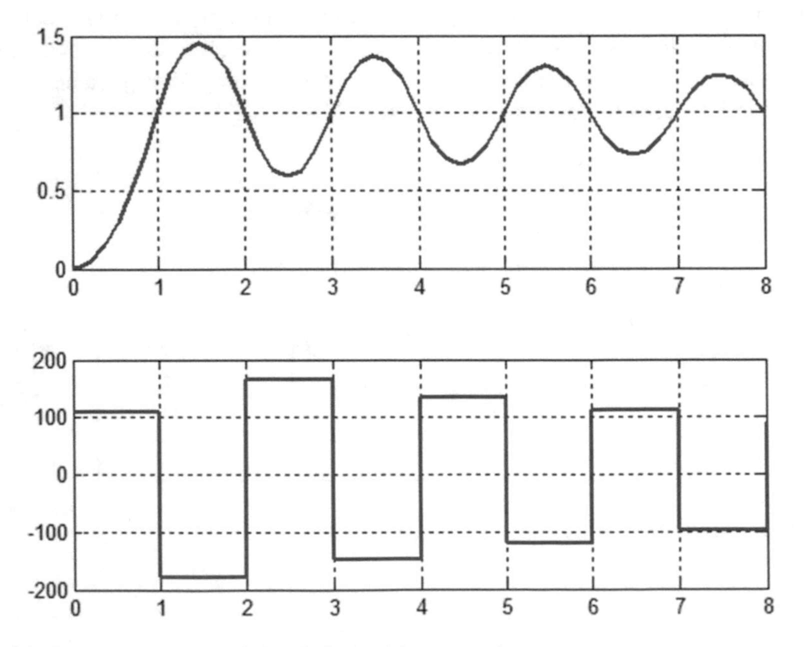

Fig. 12.17 Output and control signals in dead-beat control

written in the form $G(z) = \mathcal{B}(z)/\mathcal{A}(z)$, then the pulse transfer function of the regulator is $C(z) = \frac{\mathcal{A}(z)}{\mathcal{B}(z)(z-1)}$, and the loop transfer function is the pulse transfer function of an integrator.

$$L(z) = C(z)G(z) = \frac{\mathcal{A}(z)}{\mathcal{B}(z)(z-1)} \cdot \frac{\mathcal{B}(z)}{\mathcal{A}(z)} = \frac{1}{z-1}$$

It can be seen that the zeros of the process appear in the regulator as poles. The pulse transfer function $G(z)$ contains one zero, $z_1 = -0.9048$, which appears in the regulator as a pole. Analyse this in more detail.

```
C1z=1/(z+0.9048)
step(C1z)
```

It is seen that this component causes the oscillations. In the continuous domain as $z_1 = \mathrm{e}^{-sT_s}$, $s_1 = -\ln(\mathrm{z})/T_s$

```
s1=log(-0.9048)
  s1 = -0.1000 + 3.1416i
```

This corresponds to complex conjugate pole pairs with a small damping factor, the oscillation frequency is $\pi/T_s = 3.14$, just the twice the sampling frequency. Obviously this is not a real zero, it appeared due to sampling, therefore it does not have to be compensated. The Appendix demonstrates how the complex conjugate pole pairs are transformed to the z domain.

(As a demonstration let us analyse the step responses of the following pulse transfer functions.

```
P1z=0.5/(z-0.5); step(P1z);
P2z=1.5/(z+0.5); step(P2z);
P3z=2.5/(z+1.5); step(P3z);
```

It can be seen that the output of the first system shows an aperiodic settling process corresponding to the response of a first order lag element. The output of the second system presents decreasing oscillations, while the third system is unstable. The constants were chosen to ensure unit static gain. The regulator should not cancel the "bad" zeros of the process, i.e. those which are outside of the unit circle or lie in the unfavourable area of the unit circle. The unfavourable area, as will be seen in the Appendix, is the area outside of a "heart shape curve", which includes also the negative real zeros inside the unit circle.)

Let us separate the numerator of the pulse transfer function of the process into cancellable and non-cancellable components.

$$\mathcal{B}(z) = \mathcal{B}_+(z)\,\mathcal{B}_-(z)$$

$\mathcal{B}_+(z)$ contains the cancellable roots and $\mathcal{B}_-(z)$ those which are not cancellable. If a zero in $\mathcal{B}(z)$ is not cancelled, than it will appear in the resulting pulse transfer of the control system between the output signal and the reference signal. A requirement is to track the step reference signal without steady error. Therefore the static gain of $\mathcal{B}_{-}(z)$ should be 1, i.e. $\mathcal{B}_{-}(z)|_{z=1} = 1$.

In the next step a regulator is designed which does not compensate the zero which would cause the oscillation. In the design, the dead-time d of the process is also considered. In the sequel, the pulse transfer functions are given with the z^{-1} shift operator.

The required pulse transfer function is given in the following form:

$$T\left(z^{-1}\right) = \mathcal{B}_{-}\left(z^{-1}\right) z^{-d}$$

The pulse transfer function of the process is

$$G\left(z^{-1}\right) = \frac{\mathcal{B}_{+}(z^{-1})\mathcal{B}_{-}(z^{-1})}{\mathcal{A}(z^{-1})} z^{-d}.$$

The resulting transfer function between the output and the reference signal is

$$\frac{C(z^{-1})G(z^{-1})}{1+C(z^{-1})G(z^{-1})} = T\left(z^{-1}\right) = \mathcal{B}_{-}\left(z^{-1}\right) \cdot z^{-d}$$

Hence the pulse transfer function of the regulator:

$$C\left(z^{-1}\right) = \frac{\mathcal{B}_{-}(z^{-1}) \cdot z^{-d}}{G(z^{-1})[1 - \mathcal{B}_{-}(z^{-1}) \cdot z^{-d}]} = \frac{T(z^{-1})}{G(z^{-1})[1 - T(z^{-1})]},$$

or, considering the decomposition of $G(z)$,

$$C\left(z^{-1}\right) = \frac{\mathcal{A}(z^{-1})}{\mathcal{B}_{+}(z^{-1})[1 - \mathcal{B}_{-}(z^{-1}) \cdot z^{-d}]}$$

Apply the regulator design for our example avoiding intersampling oscillations. The separation of the numerator is

$\mathcal{B}_{-}(z^{-1}) = (1+0.9048z^{-1})/(1+0.9048)$; $\mathcal{B}_{+}(z^{-1}) = 0.0090559 \cdot (1+0.9048)$ and $d = 1$.

This is calculated by

```
Bm =1+0.9048*z^(-1)
Bmn=Bm/dcgain(Bm)
Bpn=Gz.k*dcgain(Bm)
Tz=Bmn/z
```

```
Cz=Tz/(Gz*(1-Tz))
Cz=minreal(Cz,0.001)
```

The regulator pulse transfer function is then

```
57.972(z-0.9048)(z-0.8187)
--------------------------
       (z+0.475)(z-1)
```

Simulate again the behaviour with the SIMULINK™ model. The output and the control signal are shown in Fig. 12.18.

It can be seen that there are no oscillations and at the same time the control signal became more moderate, with smaller maximum value. The control system became slower, now the output signal reaches the steady value in two sampling periods. As the overexcitation is still about 50, this value may exceed the possibilities of the applied actuator. Therefore a method has to be sought to decrease further the overexcitation while keeping the finite settling time.

Supplement the control algorithm with a *design filter polynomial*, which calmly "guides" the finite time settling process. For example, choosing the design polynomial

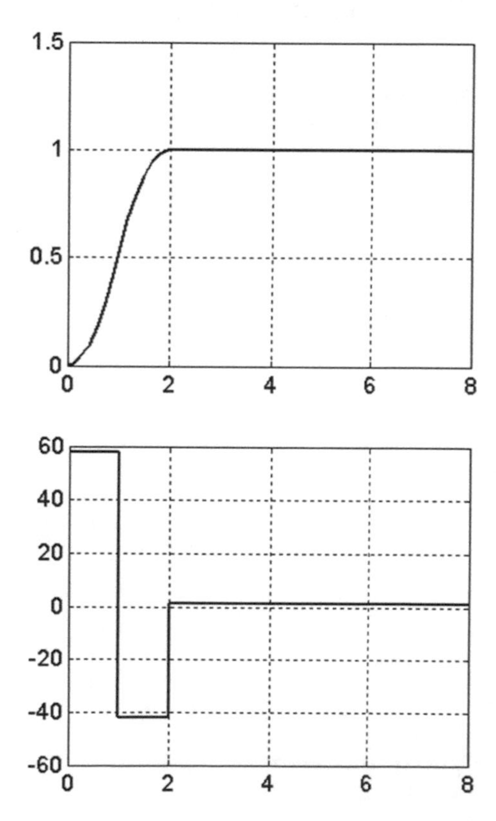

Fig. 12.18 Output and control signals in dead-beat control avoiding inter-sampling oscillations

$$\mathcal{F}(z) = \frac{1}{3} + \frac{z^{-1}}{3} + \frac{z^{-2}}{3}$$

its smoothing effect can be observed looking at its step response.

```
Fz=(1+z^(-1)+z^(-2))/3
step(Fz)
```

With the design polynomial the design criterion is

$$T\left(z^{-1}\right) = \mathcal{F}(z) \cdot \mathcal{B}_{-}\left(z^{-1}\right) \cdot z^{-d}$$

For zero static error, the static gain of the design polynomial should be 1, i.e. $\mathcal{F}(z = 1) = 1$.

The regulator is

$$C\left(z^{-1}\right) = \frac{\mathcal{F}(z^{-1})\mathcal{B}_{-}(z^{-1}) \cdot z^{-d}}{G(z^{-1})[1 - \mathcal{F}(z^{-1})\mathcal{B}_{-}(z^{-1}) \cdot z^{-d}]} = \frac{\mathcal{A}(z^{-1})\mathcal{F}(z^{-1})}{\mathcal{B}_{+}(z^{-1})[1 - \mathcal{F}(z^{-1})\mathcal{B}_{-}(z^{-1}) \cdot z^{-d}]}$$

```
Tz=Fz*Bmn/z
Cz=Tz/(Gz*(1-Tz))
Cz=minreal(Cz,0.001)
```

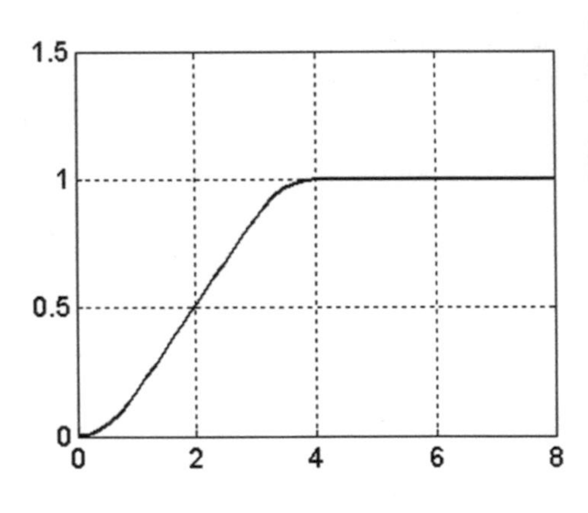

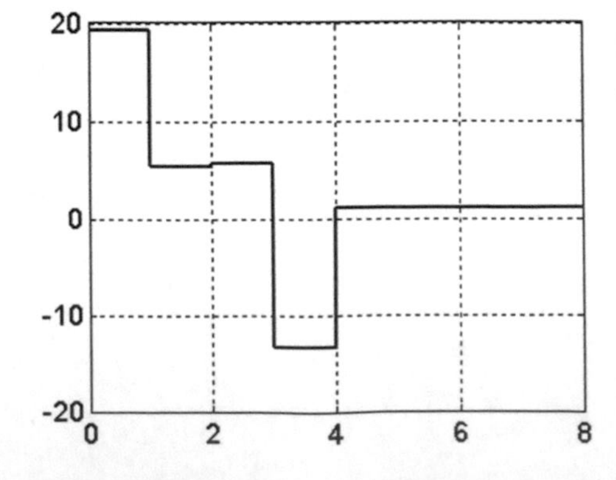

Fig. 12.19 With a design filter the settling process is modified

Running the SIMULINK™ diagram it can be seen that the settling time has been increased and the maximum value of the control signal has become lower (Fig. 12.19).

Add a dead-time $T_d = 1$ to the process and simulate the behaviour of the control system

```
Td=1
d=Td/Ts+1
Tz=Bmn/(z^d)
Cz=Tz/(Gz*(1-Tz))
Cz=minreal(Cz,0.001)
```

Put a *delay* block (Simulink –> Continuous –> Transport Delay) in the SIMULINK™ model and set its parameter to Td. Similarly to the previous discussion with a different $\mathcal{F}(z)$ design polynomial the control system could fulfill more flexible specifications.

Appendix

Let us assume that a continuous system can be characterized by a dominant pole pair, so it can be considered as a second order oscillating system. For a given damping factor ξ for different time constants the poles are located on straight lines which start from the origin of the complex plain and close angle $\cos\varphi = \xi$ with the negative real axis. With sampling these straight lines are mapped into "heart shape curves" on the complex plain. These mapping is illustrated by the following program. The location of the poles ensures acceptable transients if the damping factor is above a given value (e.g. $\xi \geq 0.6$).

The conjugate complex poles with a given damping factor in the continuous domain. In the s domain with constant damping factor ξ we get $s = \sigma + j\omega$ straight lines going through the origin, as σ takes different values. For a given value of σ, $\omega = \frac{\sigma}{\xi}\sqrt{1-\xi^2}$. In the discrete domain according to mapping $z = e^{sT_s}$ 'heart shape curves' are obtained. To demonstrate this, write the following MATLAB code:

```
szigma=0:0.01:1.6;
kszi=0.4;
Ts=1;
z=exp(Ts*(-szigma+j*sqrt(1-kszi*kszi)*szigma/kszi));
plot(real(z),imag(z),real(z),-imag(z)),grid;
```

Those roots of $\mathcal{B}(z)$ which lie inside the closed curve (where the damping is bigger than on the contour) are the roots of $\mathcal{B}_+(z)$, while those roots which lie on the curve or outside of it are the roots of $\mathcal{B}_-(z)$.

Chapter 13
Design of Discrete *PID* Regulators

In a sampled-data control circuit the regulator can be designed in the frequency domain taking into account that by sampling and applying a zero order hold the system behaves as if additional dead-time had occurred in the system.

This can be demonstrated by the following example. A first order proportional lag element is excited with a sinusoidal signal. Its output is sampled with sampling time 1 s, then a zero order hold is applied. Plot the input signal, and the outputs of both the continuous and the sampled system in the same diagram (Fig. 13.1).

The continuous output signal in quasi-stationary state is also a sinusoidal signal of the same frequency as that of the input signal, but differs from it in amplitude and phase angle. Seemingly the sampled signal is delayed compared to the continuous output signal, and its basic harmonic is delayed by about half of the sampling time.

Analyse this phenomenon in the frequency domain. Compare the frequency function of the continuous lag element with the frequency function of the sampled element with zero order hold (Fig. 13.2).

The pulse transfer function of the sampled system with $A = 1$ is

$$G(z) = \frac{1 - e^{-T_s/T_1}}{z - e^{-T_s/T_1}}$$

where T_s is the sampling time. The frequency function is obtained by substituting $z = e^{j\omega T_s}$. Approximate the exponential terms with their Taylor series:

$$G\left(z = e^{j\omega T_s}\right) = \frac{1 - e^{-T_s/T_1}}{e^{j\omega T_s} - e^{-T_s/T_1}}$$
$$\approx \frac{1 - \left(1 - \frac{T_s}{T_1} + \frac{1}{2}\left(\frac{T_s}{T_1}\right)^2 - \cdots\right)}{1 + j\omega T_s - \frac{(\omega T_s)^2}{2} + \cdots - \left(1 - \frac{T_s}{T_1} + \frac{1}{2}\left(\frac{T_s}{T_1}\right)^2 - \cdots\right)}$$

L. Keviczky et al., *Control Engineering: MATLAB Exercises*,
Advanced Textbooks in Control and Signal Processing,
https://doi.org/10.1007/978-981-10-8321-1_13

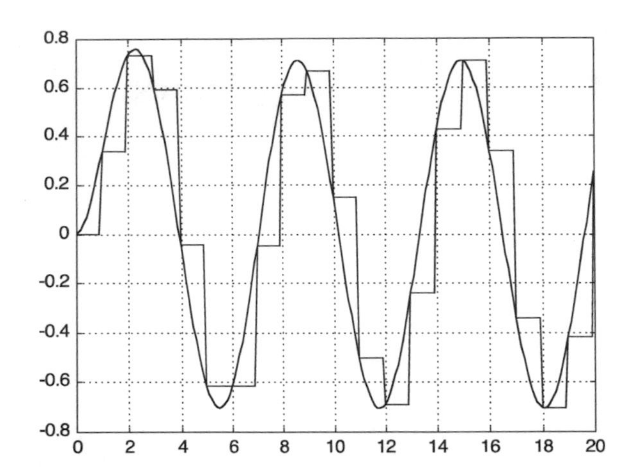

Fig. 13.1 Sampling and holding introduces an artificial additional dead time

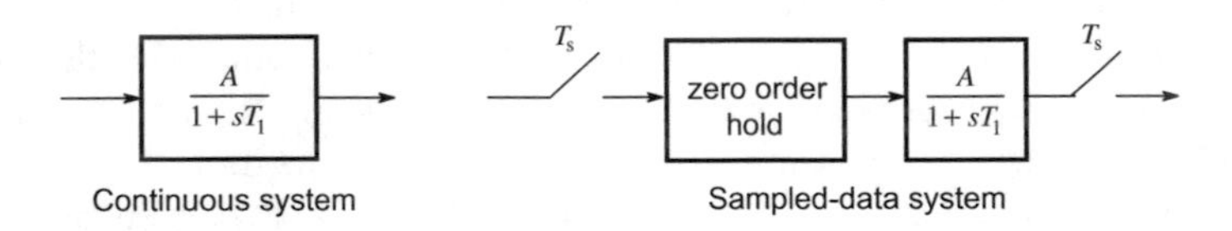

Fig. 13.2 Continuous and sampled system

If $T_s < T_1$ and $\omega T_s < 1$, the higher powers of T_s/T_1 and ωT_s can be neglected, and then the frequency function of the sampled system is approximately the same as the frequency function of the continuous system.

$$G\left(z = e^{j\omega T_s}\right) \approx \frac{T_s/T_1}{j\omega T_s + T_s/T_1} = \frac{1}{1 + j\omega T_1}$$

The neglected terms can be taken into account by an additional T_j dead-time whose value is estimated between $T_s/2$ and T_s.

$$G\left(z = e^{j\omega T_s}\right) \approx \frac{1}{1 + j\omega T_1} e^{-j\omega T_j}.$$

13.1 Comparing the Frequency Characteristics of Continuous and Discrete Systems

Compare the frequency functions of a continuous system with that of its corresponding sampled system.

Example 13.1 Let $A = 1$, $T_1 = 0.1$ and the sampling time $T_s = 0.1$.

$$P(s) = \frac{1}{1+0.1s}$$

```
s=zpk('s')
Ps=1/(1+0.1*s)
Ts=0.1;
Gz=c2d(Ps,Ts)
```

$$G(z) = \frac{0.63212}{z-0.3679}$$

```
w=logspace(-1,2,200);
[mags,phases]=bode(Ps,w);
[magz,phasez]=bode(Gz,w);
subplot(211), loglog(w, mags(:),'b',w,magz(:),'r'),grid;
PhasesWithDel=phases(:)-w'*(Ts/2)*180/pi;
subplot(212)
semilogx(w,phases(:),'b',w,phasez(:),'r',w,
PhasesWithDel,'g')
grid
```

It can be seen that the amplitude-frequency curve of the sampled system calculated from the pulse transfer function (red) follows well the amplitude-frequency curve of the continuous system (blue) up to $\omega = 1/T_s = 10$ (Fig. 13.3). At higher frequencies the deviation is high: the approximation can not be accepted. In the phase angle the deviation is observable also in the low frequency range, at $\omega = 1/T_s = 1$ it is already about 0.5 rad, nevertheless the continuous phase angle extended by the angle of the additional dead-time (green) provides a good approximation of the phase angle of the sampled system (red).

It can be seen that the amplitude of the discrete frequency function (red) beyond frequency $\omega = \pi/T_s = 31.4$ differs significantly from the amplitude of the continuous frequency function.

The additional dead-time resulting from sampling and holding changes the structural properties of the original system, even if it is very small. For instance a structurally stable continuous system will not still have this property after sampling.

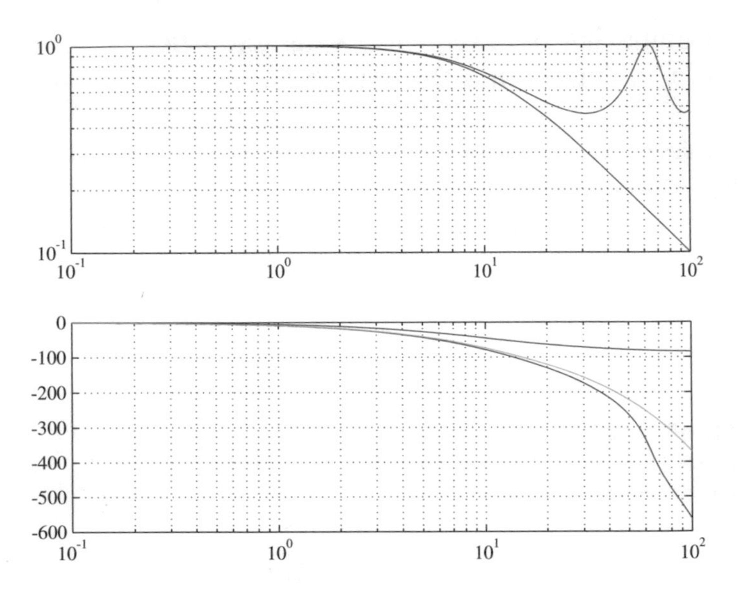

Fig. 13.3 Comparing the BODE diagrams of the continuous and the sampled systems

13.2 Design of a Discrete *PID* Regulator

The structure of a sampled control system is given in Fig. 13.4. This is a hybrid system as at some points the signals are continuous, while at others discrete signals appear.

In the figure, $P(s)$ is the transfer function of the continuous process, and $C(z)$ is the pulse transfer function of the discrete regulator to be designed. The quality specifications set for the control system prescribe the required static and dynamic properties of the system.

A continuous *PID* type regulator can be designed for the continuous process enhanced by the dead-time resulting as the effect of sampling. Then the continuous regulator is transformed to a discrete algorithm.

But it is expedient to design directly a discrete regulator of *PID* type for the pulse transfer function $G(z)$ of the process, obtained from the sampled form of the continuous process $P(s)$, which considers also the zero order hold. The design can use the pole cancellation technique. The unfavourable poles of the process are cancelled by the zeros of the regulator, and instead more favourable poles are introduced.

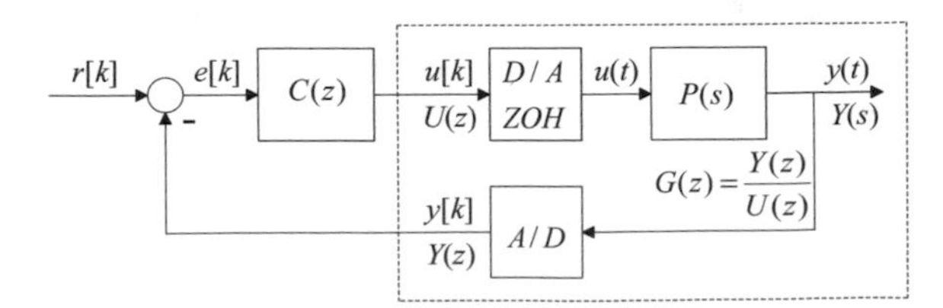

Fig. 13.4 Block diagram of a sampled control system

13.2.1 Discrete PID Regulators

The denominator of pulse transfer functions of lag elements contains factors of form $\left(z - e^{-T_s/T_1}\right)\left(z - e^{-T_s/T_2}\right)\ldots$
The discrete *P, PI, P* and *PID* algorithms can be given by the following pulse transfer functions:

$$P \text{ regulator:} \qquad C(z) = A$$

$$PI \text{ regulator:} \qquad C(z) = A\frac{z - e^{-T_s/T_1}}{z - 1}$$

(The biggest time constant can be cancelled and instead an integrating effect is introduced.)
Its difference equation is: $u[k] = Ae[k] - A\exp(-T_s/T_1)\,e[k-1] + u[k-1]$, where $u[k]$ is the control signal and $e[k]$ is the actual value of the error signal.

$$PD \text{ regulator:} \quad C(z) = A\frac{z - e^{-T_s/T_2}}{z - e^{-T_s/T_2^*}}, \quad \text{where } T_2^* < T_2.$$

(An unfavourable time constant of the process can be cancelled and instead a smaller time constant is introduced.)

$$\text{Ideal } PD \text{ regulator:} \quad C(z) = A\frac{z - e^{-T_s/T_2}}{z}$$

Its difference equation is: $u[k] = Ae[k] - A\exp(-T_s/T_1)\,e[k-1]$.

(Unlike the continuous *PD* algorithm a discrete ideal *PD* effect is realizable, as here the overexcitation is a finite value.)

$$PID \text{ regulator with ideal } PD \text{ effect:} \quad C(z) = A\frac{\left(z - e^{-T_s/T_1}\right)\left(z - e^{-T_s/T_2}\right)}{(z-1)\,z}$$

Its difference equation is:

$$\begin{aligned} u[k] = {} & Ae[k] - A(\exp(-T_s/T_1) + \exp(-T_s/T_1))\,e[k-1] \\ & + A(\exp(-T_s/T_1)\exp(-T_s/T_1))\,e[k-2] + u[k-1] \end{aligned}$$

$$PID \text{ regulator with non-ideal } PD \text{ effect:} \quad C(z) = A\frac{\left(z - e^{-T_s/T_1}\right)\left(z - e^{-T_s/T_2}\right)}{(z-1)\left(z - e^{-T_s/T_2^*}\right)}$$

Its difference equation is

$$u[k] = Ae[k] - A[\exp(-T_s/T_1) + \exp(-T_s/T_2)]e[k-1] + A[\exp(-T_s/T_1)\exp(-T_s/T_2)]e[k-2] \\ + \left[1 + \exp\left(-T_s/T_2^*\right)\right]u[k-1] - \exp\left(-T_s/T_2^*\right)u[k-2]$$

Remark: Multiple *PI* and *PD* effects can also be applied if the limit given for the control signal u is not exceeded.

The difference equation of the controller is a recursive relation which can be realized in real time.

13.2.2 Behaviour of the Basic Regulators

Example 13.2 Analyse the step responses of the individual regulators.

PI regulator:

```
Ts=1;
z=zpk('z',Ts)
Ti=10; Ts=1; A=2;
pdi=exp(-Ts/Ti)
   pdi =
       0.9048
Cpi=A*(z-pdi)/(z-1)
step(Cpi,15),grid
```

The step response is shown in Fig. 13.5.

PD regulator:

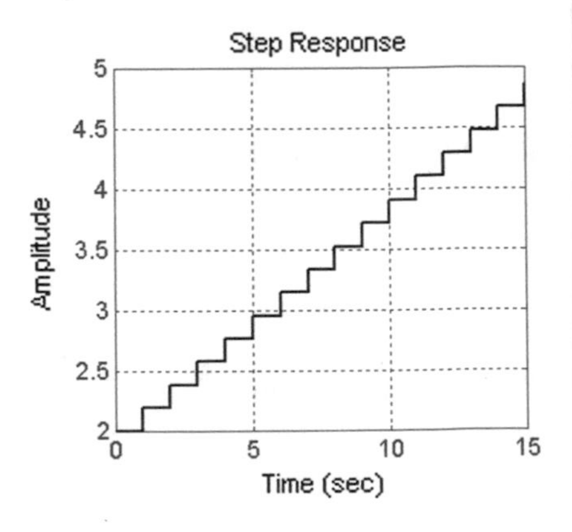

Fig. 13.5 Step response of the discrete *PI* regulator

```
Td=10;
pdd=exp(-Ts/Td)
Cpd=A*(z-pdd)/z
step(Cpd,15),grid
```

This regulator provides a big overexcitation value at the first sampling point, which results in an acceleration effect (Fig. 13.6).

If a gradually decreasing accelerating signal is required as in the continuous case then a discrete pole has also to be added analogously to the continuous case (Fig. 13.7).

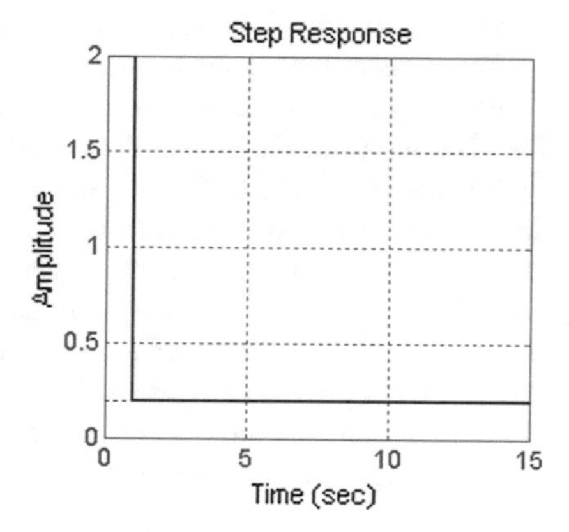

Fig. 13.6 Step response of the discrete ideal *PD* regulator

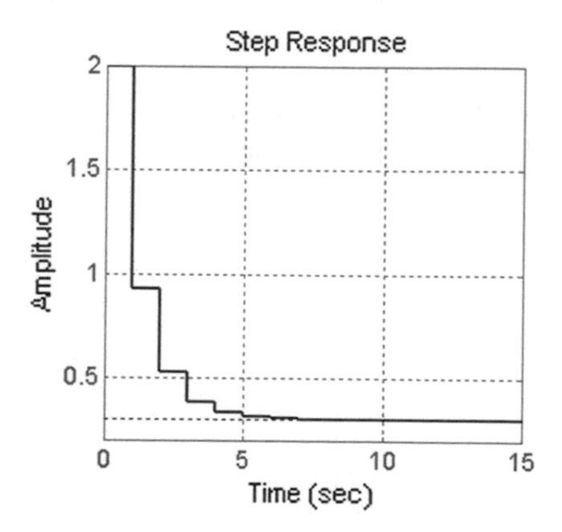

Fig. 13.7 Step response of the discrete *PD* regulator

```
Td=10;Td1=1;
pdd1=exp(-Ts/Td1)
Cpd1=A*(z-pdd)/(z-pdd1)
step(Cpd1,15),grid
```

PID regulator:

```
Cpid=Cpi*Cpd
Cpid1=Cpi*Cpd1
step(Cpid,15)
step(Cpid,Cpid1,15),grid
```

It can be seen that with the discrete *PID* regulator algorithms similar effects can be reached as with the continuous *PID* algorithms (Fig. 13.8).

The zeros of the regulator cancel the unfavourable poles of the process. The gain *A* of the regulator is chosen to ensure the required phase margin (generally $\sim 60°$) based on the BODE diagram of the discrete open loop system. The BODE diagram can be calculated by the MATLAB™ command dbode, or if the system is given in *LTI* form, then with the command bode. The frequency range has to be considered until the point $1/T_s$: the low frequency approximation of discrete systems is valid up to this value. The cut-off frequency will be obtained to be about $\omega_c \approx 1/[2\,(T_d + T_s)]$, where T_d is the dead-time of the continuous process. (From sampling the process the additional dead-time is about $T_s/2$, the regulator algorithm adds a further $T_s/2$.) Regarding regulator design, only the low frequency range is of interest, where otherwise the frequency diagram of the discrete system approximates the continuous one.

As the *PID* control algorithms do not compensate the zeros of the process, there will no intersampling oscillations. For the control signal restrictions are given. It has to be checked if the control signal exceeds the given limit or not.

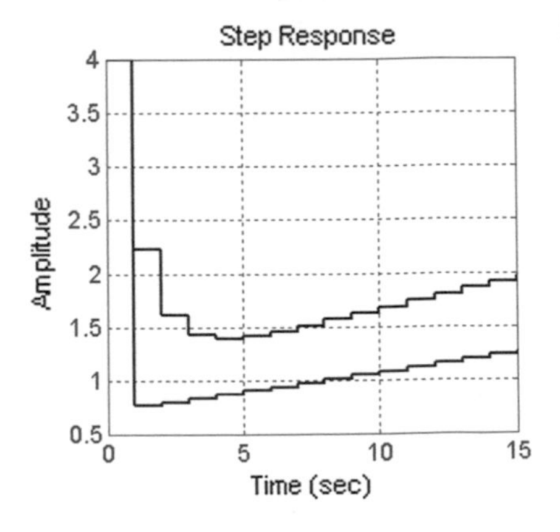

Fig. 13.8 Step response of the discrete *PID* regulator

13.2.3 Regulator Design for a Prescribed Phase Margin

Example 13.3 The continuous process is given by the transfer function

$$P(s) = \frac{e^{-s}}{(1+10s)(1+5s)} = P_{+}(s)e^{-s}$$

The system has a dead-time term of $T_d = 1$ s. The sampling time is $T_s = 1$ s.

Design a series discrete regulator to meet the following design specifications:

- Phase margin $\sim 60°$
- Settling time should be minimal
- The closed loop system should follow a unit step reference signal without steady error (type 1 system).

First define the continuous process without the dead-time ($P_{+}(s)$ = Ps):

```
s=zpk('s');
Ps=1/((1+10*s)*(1+5*s));
```

Determine the discrete process model assuming zero order hold (without the dead-time) ($G_{+}(z)$ = Gz).

$$G_{+}(z) = \left(1 - z^{-1}\right)\mathcal{Z}\left\{\frac{P_{+}(s)}{s}\right\}$$

```
Ts=1
Gz=c2d(Ps,Ts,'zoh')
```

Note that it is not necessary to include the default 'zoh' string.

Add the dead-time to the process by multiplying $G_{+}(z)$ with z^{-1}, since $\mathcal{Z}\{e^{-s}\} = z^{-1}$.

The variable z can be defined similarly to the s variable. Here the sampling time should also be given.

```
z=zpk('z',Ts)
Gz=Gz/z
```

$$G(z) = z^{-1}G_{+}(z) = \frac{1}{z}G_{+}(z) = 0.00905\frac{(z+0.9048)}{(z-0.9048)(z-0.8187)z}$$

The zeros and poles of the discrete system are

```
[zd,pd,kd]=zpkdata(Gz,'v')
```

PI compensating term is necessary to achieve the steady state zero error requirement. *PD* term is used to accelerate the system response. The *PI* and *PD* break frequencies are chosen according to the time constants (poles) of the continuous process. T_I is chosen equal to the largest time constant of the process and T_D is equal to the second largest time constant. The parameter T_D1 is determined from the given n_p pole shift ratio, $T_\mathrm{D1} = T_\mathrm{D}/n_\mathrm{p} = 5/5 = 1$. In the continuous case the regulator would be the following:

$$C(s) = k_\mathrm{c}\frac{sT_\mathrm{I}+1}{sT_\mathrm{I}}\frac{sT_\mathrm{D}+1}{sT_\mathrm{D1}+1} = k_\mathrm{c}\frac{(10s+1)(5s+1)}{10s(s+1)}.$$

The discrete equivalent of these breakpoint frequencies can be calculated on the basis of the transformation $z = e^{sT_\mathrm{s}}$ both for the zeros and the poles of the regulator:

$$PI \text{ break frequency at:} \quad 0.1 \Rightarrow \mathrm{e}^{-T_\mathrm{s}/T_\mathrm{I}} = \mathrm{e}^{-1/10} = \mathrm{e}^{-0.1} = 0.9048\,.$$

$$PD \text{ break frequency at:} \quad 0.2 \Rightarrow \mathrm{e}^{-T_\mathrm{s}/T_\mathrm{D}} = \mathrm{e}^{-1/5} = \mathrm{e}^{-0.2} = 0.8187.$$

$$PD1 \text{ break frequency at:} \quad 1 \Rightarrow \mathrm{e}^{-T_\mathrm{s}/T_\mathrm{D1}} = \mathrm{e}^{-1/1} = \mathrm{e}^{-1} = 0.3679\,.$$

The discrete regulator is

$$C(z) = k_\mathrm{c}\frac{z-0.9048}{z-1}\frac{z-0.8187}{z-0.3679}$$

The k_c parameter is calculated to set the 60° phase margin.
First assume $k_\mathrm{c} = 1$:

```
kc=1;
Cz=((z-0.9048)*(z-0.8187))/((z-1)*(z-0.3679))
```

or directly with the poles of the pulse transfer function of the process

```
Cz=((z-pd(1))*(z-pd(2)))/((z-1)*(z-exp(-1)))
```

Calculate the discrete loop pulse transfer function $L(z) = C(z)\,G(z)$.

```
Lz=Cz*Gz
Lz=minreal(Lz, 0.001)
```

The command minreal cancels a coinciding zero-pole pair, if their deviation is less than the given accuracy. If the accuracy is not given, MATLAB™ considers the preliminarily defined variable eps as an accuracy limit.

```
eps
```

For the calculation of the Bode diagram, define the following frequency vector:

```
w=logspace(-2,0,200);
```

The lower point of the frequency range is less than the reciprocal of the biggest time constant of the process, and its upper point is the reciprocal of the sampling time. The low frequency approximation is valid up to this point.

```
[mag,phase]=bode(Lz,w);
T=[mag(:), phase(:), w']
```

(The frequency vector has to be transposed in order to get a column vector.)

```
T =

    0.1350   -118.8508    0.1979
    0.1318   -119.5194    0.2026
  >>0.1287  -120.2031   0.2073<<
    0.1256   -120.9023    0.2121
    0.1226   -121.6194    0.2171
```

The value of k_c can be read from the table. It is seen that a phase shift of about $-120°$ (required by the specification of phase margin of 60°) is achieved at a magnitude of 0.1287. Consequently k_c has to be chosen as 1/0.1287 = 7.77.

```
kc=1/0.1287
```

Verify now the system behaviour. Check the value of the phase margin by the command margin.

```
Cz=kc*Cz;
Lz=kc*Lz;
figure(1)
margin(Lz);
```

Plot the output and the control signal at the sampling points for a step reference signal.

```
figure(2)
step(Lz/(1+Lz)),grid
figure(3)
step(Cz/(1+Lz)),grid
```

Let us remark that MATLAB™ calculates the values of the signals only at the sampling points. The control signal is constant between two sampling points. By running a simulation in SIMULINK™ the behaviour of the signals is obtained also between the sampling points. Then the simulation step is smaller than the sampling time.

Repeat the design using an ideal *PD* element. $C_{PD}(z) = k_c \frac{z-z_2}{z}$. Now the regulator is:

$$C(z) = k_c \frac{z - 0.9048}{z - 1} \cdot \frac{z - 0.8187}{z}$$

In this case, a phase margin of 60° can be ensured with regulator gain $k_c = 1/0.0656 = 15.24$. Repeating the simulation the output signal and the control signal are shown in Fig. 13.9. It is seen that with the non-ideal *PID* regulator the

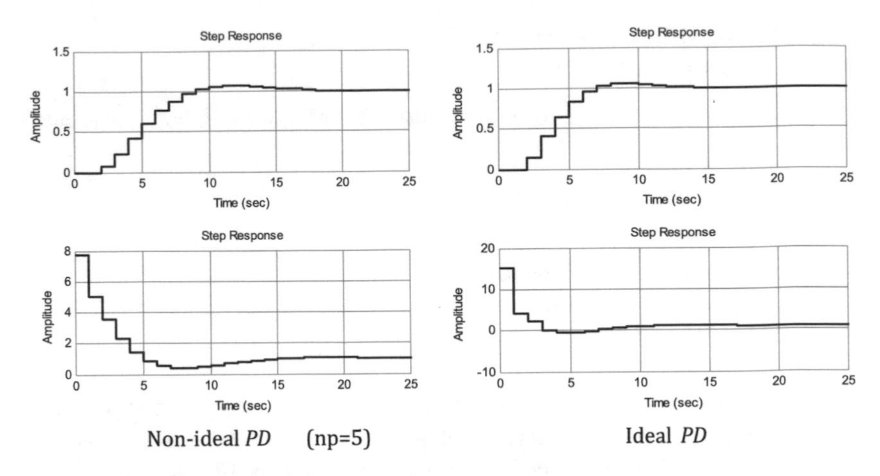

Fig. 13.9 Output and control signals of a control system with *PID* regulator

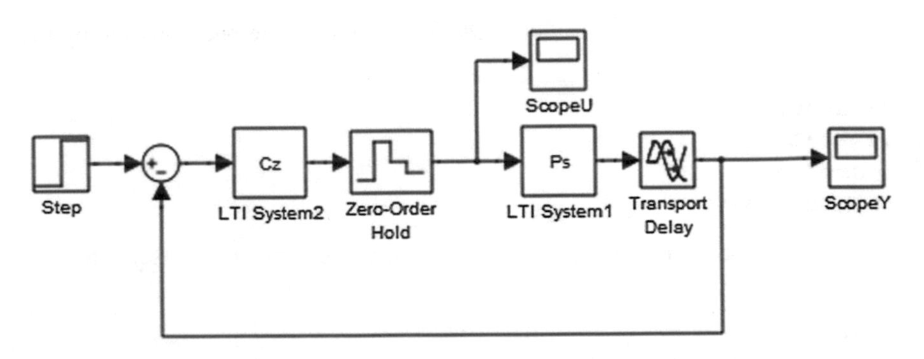

Fig. 13.10 SIMULINK™ block diagram of a sampled control system

maximum value of the control signal is smaller than in the case of the ideal *PID* regulator and the settling time is longer. The control signal starts from the value k_c. If there is a limit for the control signal, saturation can be avoided by appropriate modification of the pole shift ratio.

The closed loop performance can be investigated by the SIMULINK™ model shown in Fig. 13.10.

```
simulink
```

Create a new model (.mdl) file and copy the various blocks from the block libraries. In the upper part of the *Simulink Library Browser* window there is a *search* option. Writing here the name of the searched block, it finds it in the appropriate library, and then it can be dragged with the mouse to the model file. (*Step, Sum, LTI system, Zero-Order Hold, Transport Delay, Scope*). The blocks can be searched also in the libraries.

Step input:	*Simulink –> Sources –> Step*, *Step* time: 0
Sum:	*Simulink –> Math –> Sum*: +−
Linear system:	*Control System Toolbox –> LTI System:* Cz, Ps
Dead-time:	*Simulink –> Continuous –> Transport Delay*: Time delay: 1
Zero order hold:	*Simulink –> Discrete –> Zero-Order-Hold*: *Sampling time*: Ts
Scope:	*Simulink –> Sinks –> Scope*

The parameters of the blocks can be changed to the desired values (mouse double click).

Let us remark that the *Zero-Order-Hold* block can be omitted between the discrete regulator block and the continuous process block, as SIMULINK™ holds the output of the discrete block until the next sampling point.

The simulation can be started by menu *Start –> Simulation*, or by icon *Run* (►). The results can be visualized by double clicking the *Scope* blocks. Set the simulation time from 10 to 25.

The *Scope* block can also be used to transfer the simulation results to the MATLAB™ workspace (Fig. 13.10). Change the parameters in the *Scope* graphic window in the *Parameters* menu.

Data history:

> *Save data to workspace:* **x**
> *Variable name*: ty for the output signal and tu for the control signal.
> *Format: Array*

Running the simulation again in MATLAB™ variables ty (and tu) are created. This is a matrix, whose first column is the time vector and the second column is the output signal.

So after the simulation, the time vector t and the output vector y are obtained. From these vectors the quality measures can be determined (overshoot, settling time, maximum value of the control signal, etc.).

```
t=ty(:,1), y=ty(:,2)
```

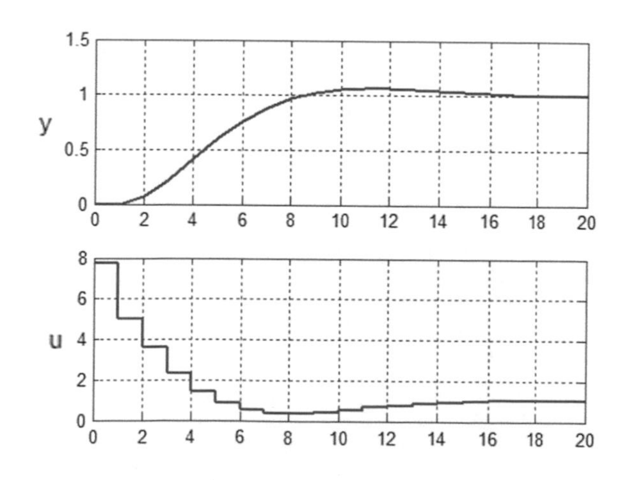

Fig. 13.11 Output and control signals in the sampled control system

Plot the output signal $y(t)$ and the control signal $u(t)$, which is the output of the hold element (Fig. 13.11). The control signal $u(t)$ is a stair-like signal, which can be plotted by using the command stairs.

```
subplot(211),plot(ty(:,1),ty(:,2)),grid
subplot(212),stairs(tu(:,1),tu(:,2)),grid
```

(Let us remark that switching the signals in the SIMULINK™ diagram to the block *To Workspace* (*Sinks* library) the data directly appear in a MATLAB™ data file.)

Chapter 14
State Feedback in Sampled Systems

Let us analyse state feedback control in sampled systems. In continuous systems the poles of the closed loop system can be set to prescribed values by feeding back the state variables of the process to the input of the process (see Chap. 9). State feedback can be applied similarly to sampled systems.

14.1 State Feedback with Pole Placement

The state equation of the sampled system is (see Chap. 11 in the textbook [1]):

$$\begin{aligned} \boldsymbol{x}[k+1] &= \boldsymbol{F}\,\boldsymbol{x}[k] + \boldsymbol{g}\,u[k] \\ y[k] &= \boldsymbol{c}^{\mathrm{T}}\boldsymbol{x}[k] + d\,u[k] \end{aligned}$$

The poles of the system in the z domain are the roots of the characteristic equation

$$\det(z\boldsymbol{I} - \boldsymbol{F}) = 0.$$

The aim of the control is the acceleration of the dynamic behaviour (or stabilization of an unstable process). One method of compensation is state feedback. Prescribing the poles of the closed loop defines the rate of acceleration.

The control signal is obtained by feedback of the discrete state variables:

$$u[k] = -\boldsymbol{k}^{\mathrm{T}}\boldsymbol{x}[k]$$

L. Keviczky et al., *Control Engineering: MATLAB Exercises*,
Advanced Textbooks in Control and Signal Processing,
https://doi.org/10.1007/978-981-10-8321-1_14

The state equation of the closed loop system is

$$\boldsymbol{x}[k+1] = \left(\boldsymbol{F} - \boldsymbol{g}\,\boldsymbol{k}^{\mathrm{T}}\right)\boldsymbol{x}[k]$$

The prescribed characteristic polynomial of the closed loop is:

$$\mathcal{R}_{\mathrm{d}}(z) = \det\left[z\boldsymbol{I} - \left(\boldsymbol{F} - \boldsymbol{g}\,\boldsymbol{k}^{\mathrm{T}}\right)\right] = (z - p_{\mathrm{d1}})\,(z - p_{\mathrm{d2}})\ldots(z - p_{\mathrm{d}n})$$

where $p_{\mathrm{d1}}, p_{\mathrm{d2}}, \ldots, p_{\mathrm{d}n}$ are the prescribed poles of the closed loop system in the z domain.

The controllability matrix of the process is

$$\boldsymbol{M}_{\mathrm{c}} = \left[\boldsymbol{g} \quad \boldsymbol{F}\boldsymbol{g} \quad \ldots \quad \boldsymbol{F}^{n-1}\boldsymbol{g}\right].$$

According to the ACKERMANN formula the state feedback vector $\boldsymbol{k}$ is calculated from the state matrices $\boldsymbol{F}$, $\boldsymbol{g}$ of the process and from the characteristic polynomial belonging to the prescribed poles of the closed loop, as follows:

$$\boldsymbol{k}^{\mathrm{T}} = [1 \quad \ldots \quad 0 \quad 0]\,\boldsymbol{M}_{\mathrm{c}}^{-1}\,\mathcal{R}_{\mathrm{d}}(\boldsymbol{F}),$$

where $\mathcal{R}_{\mathrm{d}}(\boldsymbol{F})$ is the characteristic polynomial of the closed loop by the substitution $z = \boldsymbol{F}$.

The state feedback vector $\boldsymbol{k}$ is calculated in MATLAB™ with the command acker(F, g, Rd).

```
k=acker(F,g,Rd)
```

Rd is a vector containing the prescribed poles of the feedback system [the roots of the equation $\mathcal{R}_{\mathrm{d}}(z) = 0$]. If the prescribed poles are given in continuous time, their corresponding discrete values can be determined by the transformation $z = e^{sT_{\mathrm{s}}}$, where T_{s} is the sampling time.

Example 14.1 The continuous process is given by a third order lag element. Its transfer function is

$$P(s) = \frac{6}{(s+1)\,(s+2)\,(s+3)} = \frac{1}{(1+s)\,(1+0.5s)\,(1+0.333s)}$$

(a) Give the continuous state equations of the process, then with sampling time $T_{\mathrm{s}} = 0.2$ determine the discrete state equation supposing zero order hold.
(b) Design a state feedback control, prescribing the poles of the discrete closed loop. The poles are given in continuous time, then they are transformed to discrete time with the transformation $z = e^{sT_{\mathrm{s}}}$.
Let the prescribed continuous poles be $\mathsf{Rc} = [-6 \quad -3+4j \quad -3-4j]$.
(c) Analyse the behaviour of the system for initial conditions, and for reference signal tracking and disturbance rejection.

Solution The state equation of the continuous process:

```
po=[-1 -2 -3]
[A,b,c,d]=tf2ss(6,poly(po))
Hc=ss(A,b,c,d)
    A =
       -6  -11   -6
        1    0    0
        0    1    0
    b =
        1
        0
        0
    c =
        0    0    6
    d =
        0
```

The sampling time:

```
Ts = 0.2
```

Transformation to discrete state equation:

```
Hd=c2d(Hc,Ts,'zoh')
[F,g,cd,dd]=ssdata(Hd)
        F =
              0.1977     -1.2693     -0.6483
              0.1081      0.8461     -0.0807
              0.0135      0.1888      0.9940
        g =
              0.1081
              0.0135
              0.0010
        cd =
               0          0          6
        dd =   0
```

The prescribed continuous poles for the closed loop system:

```
Rc=[-6; -3+i*4; -3-i*4]
```

Their corresponding discrete values with the given sampling time:

```
Rd=exp(Rc*Ts)
```

The obtained values are

```
Rd =
    0.3012
    0.3824 + 0.3937i
    0.3824 - 0.3937i
```

Apply the ACKERMANN formula to determine the state feedback vector.

```
k=acker(F,g,Rd)
k =
      4.2463    32.4319    77.4220
```

The parameter matrices of the state equation of the closed loop:

```
Fc=F-g*k; gc=g;cc=cd;dc=dd;
Tk=ss(Fc,gc,cc,dc,Ts)
```

The static gain:

```
kr=1/dcgain(Tk)
kr = 13.9037
```

The state parameter matrices of the closed-loop compensated with the static gain:

```
Fck=Fc;gck=kr*gc;cck=cc;dck=dc;
Tk1=ss(Fck,gck,cck,dck,Ts)
```

The step response of the closed loop (Fig. 14.1.):

```
step(Tk1,3)
```

Remark The static gain can be calculated also according to the following considerations. With state feedback we would like to arrange that for a step reference signal r, the output signal y in steady state is equal to the constant value of the reference signal. Then the derivatives of the state variables are zeros. The reference signal acts on the input of the control system through the correction factor k_r. The relation is given for the case of a *single input—single output* (*SISO*) system. In steady state the values of the state variables at the sampling point $n+1$ is the same

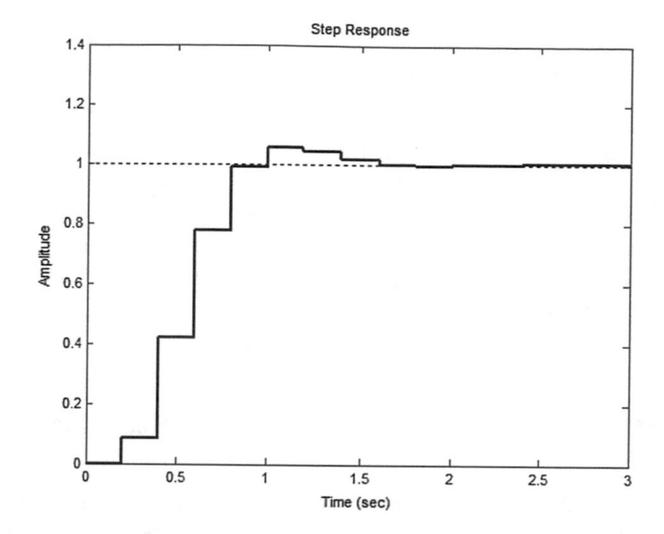

Fig. 14.1 Step response of the sampled control system

as at the point n, and the steady value of the output signal is the same as the reference signal.

$$\begin{aligned} \boldsymbol{x}_\infty &= \boldsymbol{F}\boldsymbol{x}_\infty + \boldsymbol{g}\,u_\infty \\ u_\infty &= k_r r - \boldsymbol{k}\boldsymbol{x}_\infty \\ y_\infty &= \boldsymbol{c}^T\boldsymbol{x}_\infty = r; \quad r \neq 0 \end{aligned}$$

whence

$$\begin{aligned} \boldsymbol{x}_\infty &= \left(\boldsymbol{I} - \boldsymbol{F} + \boldsymbol{g}\boldsymbol{k}^T\right)^{-1}\boldsymbol{g}\,k_r r \\ y_\infty &= \boldsymbol{c}^T\left(\boldsymbol{I} - \boldsymbol{F} + \boldsymbol{g}\boldsymbol{k}^T\right)^{-1}\boldsymbol{g}\,k_r r = r \end{aligned}$$

and the correction factor is expressed as

$$k_r = 1 \Big/ \left[\boldsymbol{c}^T\left(\boldsymbol{I} - \boldsymbol{F} + \boldsymbol{g}\boldsymbol{k}^T\right)^{-1}\boldsymbol{g}\right]$$

In our example

```
kr=1/(cd*inv(eye(3)-F+g*k)*g)
        kr = 13.9037
```

The result is the same as obtained before.

Problem Build the SIMULINK™ diagram of the control system. Analyse the behaviour of the system for initial conditions, for step reference signal and for output disturbance. Analyse the behaviour between the sampling points also.

Remark The static compensation ensures the accurate tracking of the reference signal in steady state, but does not eliminate the static error of disturbance rejection. To ensure this the state model should be enhanced with the state variables of the disturbance, and state feedback should be designed again for the enhanced system. Another possibility is extension of the system with an integrator and designing state feedback to the enhanced system.

14.2 State Feedback with Extension with Integrator

In order to track accurately the step reference signal and to decrease the effect of the disturbance it is expedient to include an integrator in the control circuit. Extend the state space model of the process with an additional state variable which is the integral of the output signal (Fig. 14.2). (A discrete equivalent of the system given in Fig. 9.6 is created.)

The difference equation of the integrator is

$$x_{\mathrm{i}}[k+1] = x_{\mathrm{i}}[k] + T_{\mathrm{s}} y[k] = x_{\mathrm{i}}[k] + T_{\mathrm{s}}\,\boldsymbol{c}^{\mathrm{T}}\boldsymbol{x}[k].$$

The extended state equation is

$$\begin{bmatrix} \boldsymbol{x}[k+1] \\ x_{\mathrm{i}}[k+1] \end{bmatrix} = \begin{bmatrix} \boldsymbol{F} & 0 \\ T_{\mathrm{s}}\,\boldsymbol{c}^{\mathrm{T}} & 1 \end{bmatrix} \begin{bmatrix} \boldsymbol{x}[k] \\ x_{\mathrm{i}}[k] \end{bmatrix} + \begin{bmatrix} \boldsymbol{g} \\ 0 \end{bmatrix} u[k] = \boldsymbol{F}_{\mathrm{b}}\boldsymbol{x}_{\mathrm{b}}[k] + \boldsymbol{g}_{\mathrm{b}}\,u[k]$$

$$y[k] = \begin{bmatrix} \boldsymbol{c}^{\mathrm{T}} & 0 \end{bmatrix} \begin{bmatrix} \boldsymbol{x}[k] \\ x_{\mathrm{i}}[k] \end{bmatrix} + d\,u[k] = \boldsymbol{c}_{\mathrm{b}}^{\mathrm{T}}\boldsymbol{x}_{\mathrm{b}}[k] + d\,u[k]$$

(The indices b refer to the extended state variables and parameter matrices.)

The state feedback of the extended system can be built according to Fig. 9.7, with the discrete system and the discrete integrator. As the number of the state variables has been increased by one, the number of the prescribed poles has to be also increased by one. The state feedback vector $\boldsymbol{k}_{\mathrm{b}}^{\mathrm{T}}$ which ensures the prescribed poles $\boldsymbol{p}_b$, the roots of $\det\left(s\boldsymbol{I} - \boldsymbol{F}_{\mathrm{b}} + \boldsymbol{g}_{\mathrm{b}}\boldsymbol{k}_{\mathrm{b}}^{\mathrm{T}}\right) = 0$, is calculated by the ACKERMANN formula with the extended parameter matrices $\boldsymbol{F}_{\mathrm{b}}$ and $\boldsymbol{g}_{\mathrm{b}}$.

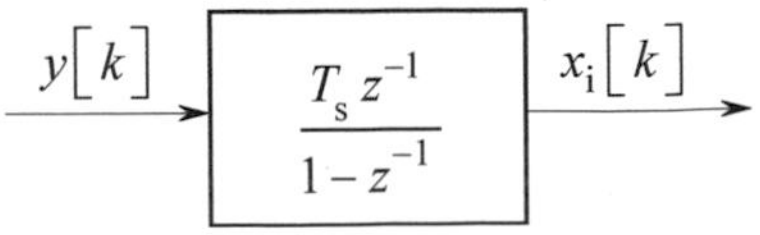

Fig. 14.2 Discrete integrator

The vector $\boldsymbol{k}_e^T$ containing the first n elements of $\boldsymbol{k}_b^T$ gives the state feedback coefficients of the original state variables. The constant k_i which gives the feedback of the integrator is the last, $n+1$-th element of $\boldsymbol{k}_b^T$.

For an *SISO* system supposing $d = 0$ the state equation of the *closed loop* system can be given by the following vector-matrix equation:

$$\begin{bmatrix} \boldsymbol{x}[k+1] \\ x_i[k+1] \end{bmatrix} = \begin{bmatrix} \boldsymbol{F} - \boldsymbol{g}\boldsymbol{k}_e^T & \boldsymbol{g}\,k_i \\ -T_s\boldsymbol{c}^T & 1 \end{bmatrix} \begin{bmatrix} \boldsymbol{x}[k] \\ x_i[k] \end{bmatrix} + \begin{bmatrix} \boldsymbol{0} \\ T_s \end{bmatrix} r[k]$$

$$y[k] = \begin{bmatrix} \boldsymbol{c}^T & 0 \end{bmatrix} \begin{bmatrix} \boldsymbol{x}[k] \\ x_i[k] \end{bmatrix} + 0 \cdot r[k]$$

Example 14.2 Let us extend the system given in Example 14.1 with an integrator and realize state feedback to the extended system. Analyse the course of the output signal for step reference input signal.

The MATLAB™ program is

```
clear
clc
po=[-1 -2 -3]
[A,b,c,d]=tf2ss(6,poly(po));
Hc=ss(A,b,c,d);
Ts=0.2
% The discrete state equation
Hd=c2d(Hc,Ts,'zoh');
[F,g,cd,dd]=ssdata(Hd)
% Extension by integrator
Fb=[F zeros(3,1);Ts*cd 1];
gb=[g;0];
cb=[c 0];
Hdb=ss(Fb,gb,cb,d,Ts)
% The prescribed poles
Rb=[-9 -6 -3+i*4 -3-i*4]
Rd=exp(Rb*Ts)
% The state feedback vector
k=acker(Fb,gb,Rd)
kk=k(1:3);
ki=k(4)
% State equation of the closed loop
Fc=[F-g*kk g*ki;-Ts*c 1]
gc=[zeros(3,1);Ts]
cc=cb
dc=0
Hdc=ss(Fc,gc,cc,dc,Ts)
step(Hdc)
```

The extended state equation:

```
Fb = 0.1977   -1.2693   -0.6483        0
     0.1081    0.8461   -0.0807        0
     0.0135    0.1888    0.9940        0
          0         0    1.2000   1.0000
gb = 0.1081
     0.0135
     0.0010
          0
cb = 0    0    6    0
```

The state feedback vector:

```
k = 6.7401      60.9170     260.8297      58.0271
```

The step response of the closed loop is shown in Fig. 14.3.

The SIMULINK™ block diagram of the control system extended with the integrator is shown in Fig. 14.4. The state equation block is taken from the *discrete* block library. The state variables have to be measurable. In the figure the setting of

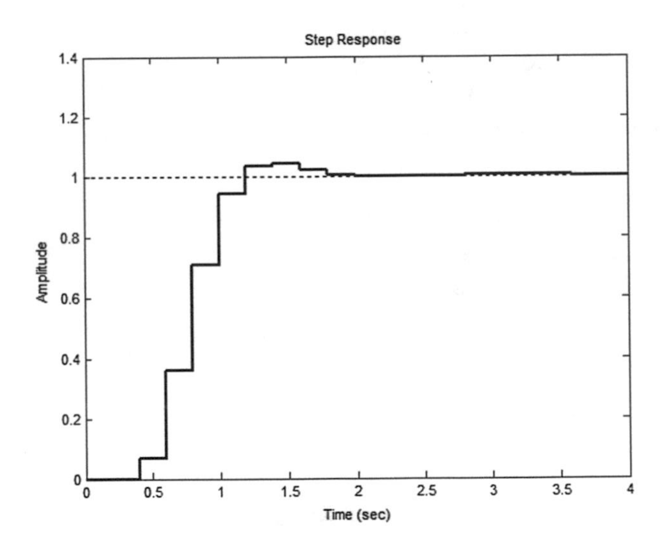

Fig. 14.3 Step response of the sampled state feedback control system extended with integrator

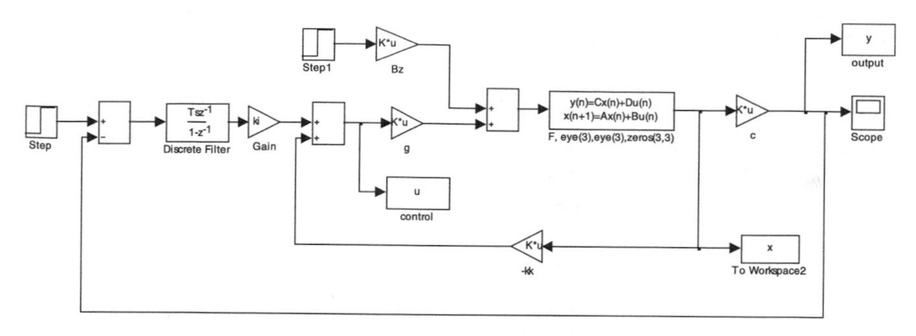

Fig. 14.4 SIMULINK™ block diagram of the discrete control system

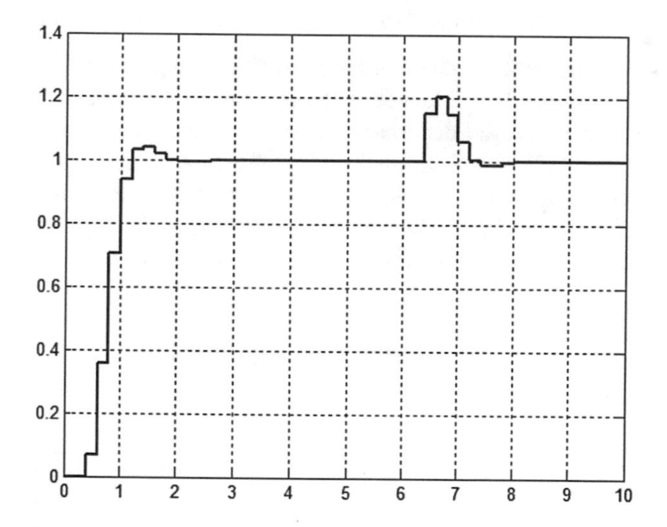

Fig. 14.5 Time course of the output signal

the parameters is indicated. Simulate the behaviour of the system by running the SIMULINK™ program for the initial conditions, for a step reference signal and for a disturbance acting at the input of the process. Let `Bz=[0 1 0]'`. In Fig. 14.5 it can be seen, that the control system tracks the reference signal without steady error, and rejects the effect of the step disturbance of amplitude 0.2 acting at the time point t = 6 s.

Problem Supplement the SIMULINK™ model with the state space model of the continuous process. Analyse the course of the output signal also between the sampling points.

14.3 State Estimation

If the state variables are not measurable, they have to be estimated. The observer can be applied for state estimation. The discretized form of the circuit shown in Fig. 9.9 is realized. If the process is known, its model is built. Figure 14.6 shows the SIMULINK™ block diagram of the state estimation of the discrete system.

The process and its model are excited by the same input signal. Comparing the output signals of the process and the model an error signal is obtained which is used to set the state variables of the model through the parameter $\boldsymbol{l}$ (see textbook [1]). The values of the estimated state variables will approach quickly and follow the values of the real state variables, if the dynamics of the estimation circuit is much faster than the dynamics of the process. The poles of the estimation circuit can be prescribed and then applying the ACKERMANN formula vector $\boldsymbol{l}$ can be determined.

Example 14.3 Consider the process of the proportional system with three time lags investigated in Example 14.1. The initial values of all the three state variables are 1. The reference signal and the disturbance signal are zero. Suppose the prescribed poles of the estimation circuit are real and of the same value, and ensure faster transients as the smallest time constant of the state feedback system (in case of conjugate complex poles let us consider the reciprocal of the absolute value).

Prescribe the poles of the continuous closed loop system:

```
Rc=[-6; -3+i*4; -3-i*4]
```

Set the poles of the continuous state estimation circuit:

```
Fc=[-7 -7 -7]
```

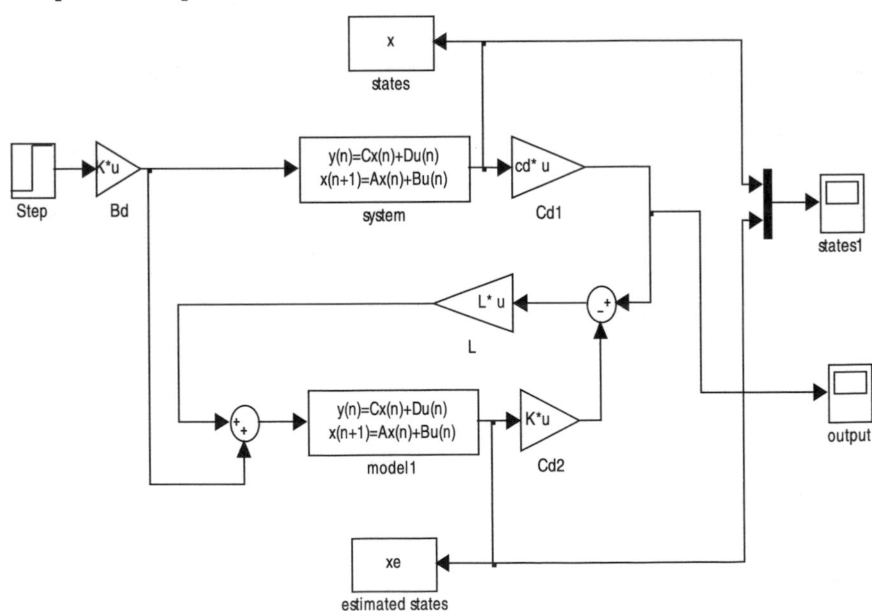

Fig. 14.6 SIMULINK™ diagram for state estimation in discrete system

The poles of the discrete state estimation circuit are

```
Fd=exp(Ts*Fc)
```

The parameters of the estimation circuit (the elements of vector l, in MATLAB™ L) in the discrete system are determined by the command

```
L=acker(F',c',Fd)'
```

The MATLAB™ program for the discrete version of the algorithm of Fig. 9.10 is

```
po=[-1,-2,-3]
[A,b,c,d]=tf2ss(6,poly(po))
Hc=ss(A,b,c,d)
Ts=0.2
Hd=c2d(Hc,Ts,'zoh')
[F,g,cd,dd]=ssdata(Hd)
% prescribed poles of the continuous estimation
Pc=[-7 -7 -7]
% poles of the discrete state estimation
Pd=exp(Pc*Ts)
% parameters of estimation circuit (elements of vector L)
L=acker(F',cd',Pd)'
Fest=F-L*cd
sysest=ss(Fest,L,cd,dd,Ts)
x0=[1;1;1]
t=0:Ts:6;
[y,t,x]=initial(Hd,x0,t);
figure(1)
stairs(t,x),grid
x0est=[0;0;0]
[yest,t,xest]=lsim(sysest,y,t,x0est);
figure(2)
stairs(t,xest),grid
figure(3)
stairs(t,x(:,1))
hold
stairs(t,xest(:,1)),grid
figure(4)
stairs(t,y),grid
hold
stairs(t,yest)
```

The state estimation vector:

```
L' = -0.7709    0.2062    0.2163
```

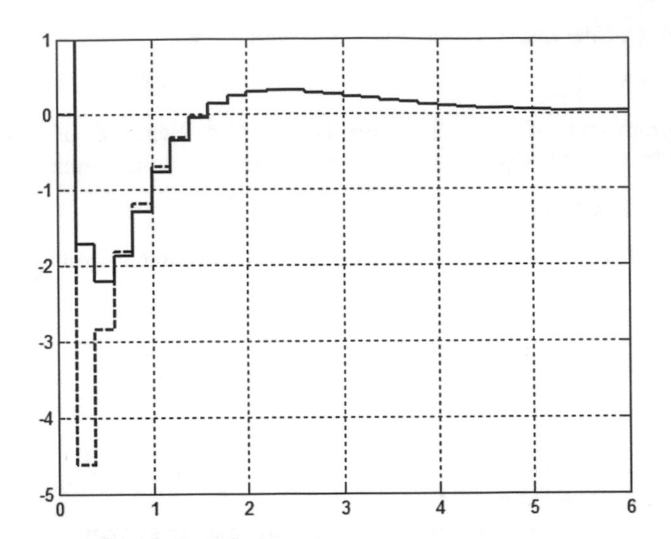

Fig. 14.7 The time course of the real and estimated first state variables

The simulation shows the fast settling of the state variables. Plot the real and the estimated first state variable in the same diagram (Fig. 14.7).

Running the SIMULINK™ model yields a similar result.

Problem Supplement the SIMULINK™ model with the state space model of the continuous process. Analyse the course of the output signal also between the sampling points.

14.4 State Feedback from the Estimated State Variables

State feedback control can be realized from the estimated state variables with the same feedback constants that are calculated for feedback from the original, real state variables (the separation principle). The control system operates well if the prescribed poles of the estimation circuit ensure faster behaviour of the estimation circuit than that of the feedback control circuit. The SIMULINK™ diagram of state feedback from the estimated state variables is shown in Fig. 14.8.

Example 14.4 The proportional process with three time lags given in Example 14.1. is sampled with sampling time `Ts=0.2`. The initial value of all the three state variables is 1. The state variables are estimated, then the control signal is

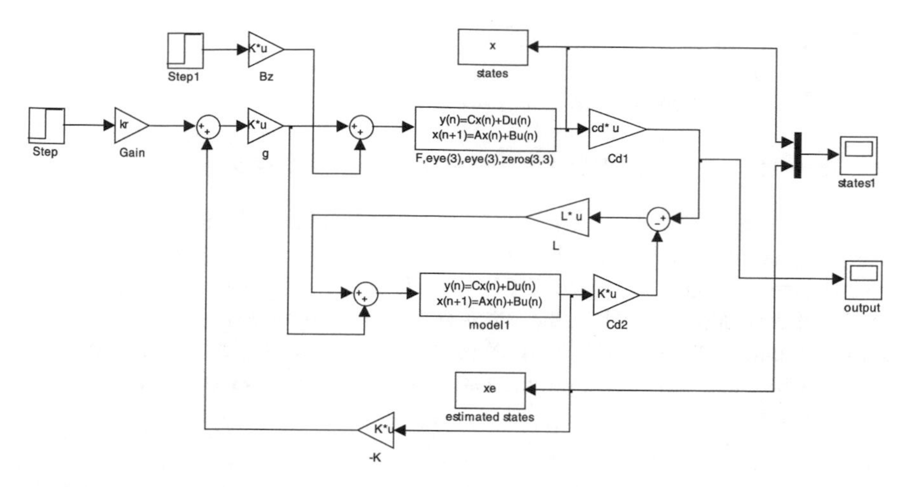

Fig. 14.8 Discrete state feedback from the estimated state variables

produced by feeding back the estimated state variables. Static compensation is applied. The prescribed poles for estimation in the continuous system are set by **`Fc=[-7-7 -7]`**. The prescribed poles for the closed loop continuous system are set by **`Rc=[-6; -3+i*4; -3-i*4]`**. The reference signal is a unit step. Let us simulate the output signal.

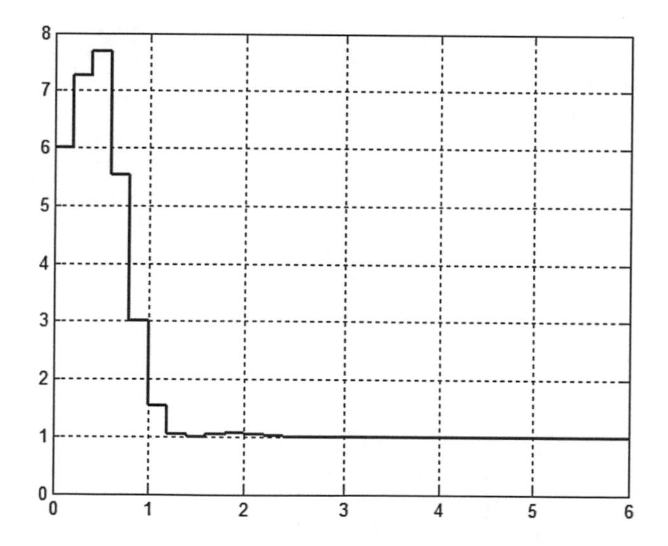

Fig. 14.9 Settling of the output signal for step reference signal and initial conditions

The values of the parameters are:

```
kr=13.9037; k=4.2463    32.4319    77.422;
L=-0.7709 0.2062    0.2163
```

The result of running SIMULINK™ is shown in Fig. 14.9.

Problem Compare the result with the time response obtained for the case without state estimation. Analyse the behaviour of the system if there is also an input disturbance. Extend the SIMULINK™ block scheme with the continuous state model of the process. Analyse the behaviour of the output signal also between the sampling points. Write a MATLAB™ program for analysing the behaviour of the system with state feedback from the estimated state variables (for the continuous case see Example 9.5).

Chapter 15
General Polynomial Method to Design Discrete Regulators

Polynomial design is an important method of regulator design. The main idea is that the transfer function of the closed loop control system is prescribed as the aim of the control, then the regulator is calculated using the knowledge of the process. The principle seems simple, nevertheless the calculation of the regulator necessitates the solution of a polynomial (Diophantine) equation, whose solvability sometimes requires the fulfilment of complicated conditions. Handling the unstable poles and inverse unstable zeros of the process requires further considerations. In Chap. 10, the design method was shown for continuous systems, and some examples demonstrated its application. For continuous systems the method can be applied only to systems without dead-time. For discrete (sampled) systems, polynomial design can be applied also to systems with dead-time.

Let the pulse transfer function of the process be

$$G(z^{-1}) = \frac{\mathcal{B}}{\mathcal{A}} z^{-d} = \frac{\mathcal{B}_+ \mathcal{B}_-}{\mathcal{A}_+ \mathcal{A}_-} z^{-d} = \left(\frac{\mathcal{B}_+}{\mathcal{A}_+}\right)\left(\frac{\mathcal{B}_-}{\mathcal{A}_-}\right) z^{-d} = G_+ G_- z^{-d}.$$

Here $\mathcal{A}_+$ contains the stable, while $\mathcal{A}_-$ contains the unstable poles of the process. $\mathcal{B}_+$ contains the stable, compensable, while $\mathcal{B}_-$ contains the unstable, non compensable zeros.

The pulse transfer function of the regulator is sought in the following form:

$$C(z^{-1}) = \frac{\mathcal{Y}}{\mathcal{X}} = \frac{\mathcal{A}_+ \mathcal{Y}_\mathrm{d} \mathcal{Y}'}{\mathcal{B}_+ \mathcal{X}_\mathrm{d} \mathcal{X}'}$$

Here $\mathcal{Y}_\mathrm{d}$ and $\mathcal{X}_\mathrm{d}$ are given polynomials. In the design the polynomials $\mathcal{Y}'$ and $\mathcal{X}'$ have to be determined so as to ensure that the poles of the closed loop control system have the prescribed values.

L. Keviczky et al., *Control Engineering: MATLAB Exercises*,
Advanced Textbooks in Control and Signal Processing,
https://doi.org/10.1007/978-981-10-8321-1_15

The characteristic equation is

$$1 + CG = 1 + \frac{\mathcal{A}_+ \, \mathcal{Y}_\mathrm{d} \mathcal{Y}'}{\mathcal{B}_+ \, \mathcal{X}_\mathrm{d} \mathcal{X}'} \frac{\mathcal{B}_+ \mathcal{B}_-}{\mathcal{A}_+ \mathcal{A}_-} z^{-d} = 0$$

or

$$\mathcal{X}_\mathrm{d}\mathcal{X}'\mathcal{A}_- + \mathcal{Y}_\mathrm{d}\mathcal{Y}'\mathcal{B}_- z^{-d} = \mathcal{R} = 0.$$

The roots of the characteristic polynomial $\mathcal{R}$ are the prescribed values. For stable behaviour they have to be located inside the unit circle.

Choose the values $\mathcal{Y}_\mathrm{d} = 1$ and $\mathcal{X}_\mathrm{d} = 1$. Let us remark that if the polynomial $\mathcal{X}_\mathrm{d} = 1 - z^{-1}$ is chosen, then an integrator is introduced into the regulator.

Suppose that the degrees of the numerator and of the denominator of the regulator are the same. Let the degree of the denominator of the regulator be less by 1 than the degree of the denominator of the process. The numerator and denominator of the regulator are obtained by solving the Diophantine equation.

Let us consider the MATLAB™ simulation of some examples discussed in this chapter of the textbook [1].

Example 15.1 The pulse transfer function of the unstable process is: $G = \frac{\mathcal{B}}{\mathcal{A}} = \frac{-0.2}{z-1.2}$.

Find a regulator in the form $C = \mathcal{Y}/\mathcal{X}$ which stabilizes the process through prescribing the characteristic polynomial $\mathcal{R} = z - 0.2 = 0$. The regulator is supposed to be of order $n - 1 = 0$. $C = \mathcal{Y}/\mathcal{X} = K/1$. Solving the characteristic equation $\mathcal{A}\mathcal{X} + \mathcal{B}\mathcal{Y} = \mathcal{R}$,

$$(z - 1.2) - 0.2K = z - 0.2.$$

The regulator is $C = K = -5$. (Let us remark that here $\mathcal{Y}_\mathrm{d} = 1$ and $\mathcal{X}_\mathrm{d} = 1$ were chosen.)

Simulate the behaviour of the closed loop with MATLAB™:

```
z=zpk('z');
Gz=-0.2/(z-1.2);
Cz=-5;
Lz=Gz*Cz;
Lz=minreal(Gz*Cz);
Tz=Lz/(1+Lz);Tz=minreal(Tz)
subplot(211),step(Tz,10); grid
subplot(212),
step(Cz/(1+Lz),10);grid
```

The step response of the controlled output signal and the regulator output are shown in Fig. 15.1. The closed loop is stable but there is a static error. This can be compensated by a prefilter, or by an extra integrator in the regulator design phase.

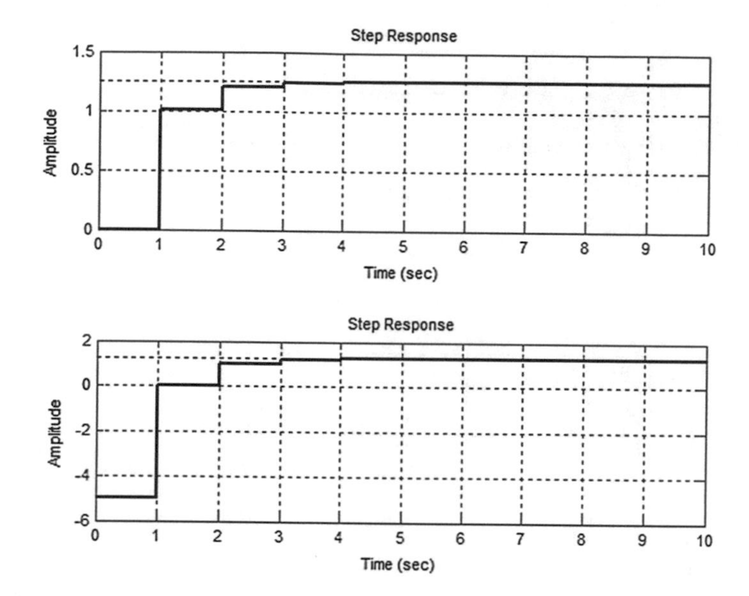

Fig. 15.1 Output and control signals for step reference input

In this system there are only discrete time pulse transfer functions, therefore the simulation can be executed without specifying the sampling time.

Example 15.2 (Example 15.3 in the textbook [1])
The pulse transfer function of the unstable process containing also dead-time is:

$$G(z^{-1}) = \frac{\mathcal{B}(z^{-1})}{\mathcal{A}(z^{-1})} = \frac{-0.2z^{-1}}{1-1.2z^{-1}}z^{-1} = \frac{-0.2}{z(z-1.2)}$$

Find a stabilizing regulator $C = \mathcal{Y}/\mathcal{X}$ prescribing the characteristic polynomial $\mathcal{R}(z) = (1-0.2z^{-1})^2$. Formally the process is of second order, therefore a characteristic polynomial of second degree is chosen. The regulator is chosen to be of first degree.

$$C = \frac{\mathcal{Y}}{\mathcal{X}} = \frac{y_o}{1+x_o z^{-1}} = \frac{y_o z}{z+x_o}.$$

The DIOPHANTINE equation is

$$(1-1.2z^{-1})(1+x_o z^{-1}) - 0.2y_o z^{-2} = (1-0.2z^{-1})^2.$$

Its solution yields $y_o = -5$ and $x_o = 0.8$. So the regulator is

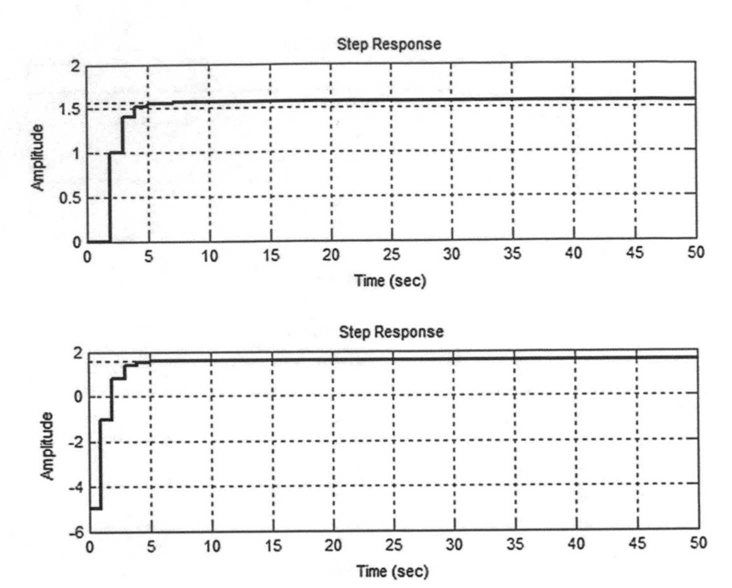

Fig. 15.2 Output and control signals for step reference input

$$C(z^{-1}) = \frac{-5}{1 + 0.8z^{-1}} = \frac{-5z}{z + 0.8}.$$

The MATLAB™ simulation results from the following code:

```
z=zpk('z');
Gz=-0.2/z/(z-1.2);
Cz=-5*z/(z+0.8)
Lz=Gz*Cz; Lz=minreal(Lz);
Tz=Lz/(1+Lz);Tz=minreal(Tz);
subplot(211),step(Tz,50),grid
Uz=Cz/(1+Lz)
subplot(212),step(Uz,50),grid
```

Figure 15.2 shows the output and the control signals for a unit step reference signal. The control system is stable, but there is a static error.

Problem Introduce an integrator in the regulator. Write the DIOPHANTINE equation and determine the parameters and the pulse transfer function of the regulator.

Chapter 16
Case Study

When analysing a process its model has to be established so that it describes correctly the static and dynamic behaviour of the outputs as responses to the given inputs and gives also the evolution in time of the state variables as the response to the initial values and the inputs.

To build the model the real operation of the system is taken into account. It has to be understood as deeply as possible. One analyses in which operating range of the input signal can the system be considered linear, or in which range of the operating points can it be linearized. The real operation is described then by mathematical equations (generally by differential equations or state equations). The values of the parameters in the equations have to be given. The parameters are known or have to be determined by measurements or by identification. Identification is a procedure where the values of the parameters are estimated from the data of input-output measurements.

On the basis of the model the output signals and the state variables of the system, as responses to the input signals and initial conditions can be calculated or simulated. Based on the model a regulator can be designed for the system to fulfil the quality specifications.

In the sequel the process of establishing the model of a heating process will be discussed (see also Sect. 2.6. in the textbook [1]).

16.1 Modelling and Analysing a Heat Process

Let us analyse the heat process of a system consisting of two heat sources. The arrangement is shown in Fig. 16.1. The temperature changes in several pieces of electrical equipment can be modelled by analysing the warming processes in two embedded bodies [5]. For example the warming processes in slots of electrical machines, where the copper winding is placed in the iron slots can be analysed on the basis of this model.

L. Keviczky et al., *Control Engineering: MATLAB Exercises*,
Advanced Textbooks in Control and Signal Processing,
https://doi.org/10.1007/978-981-10-8321-1_16

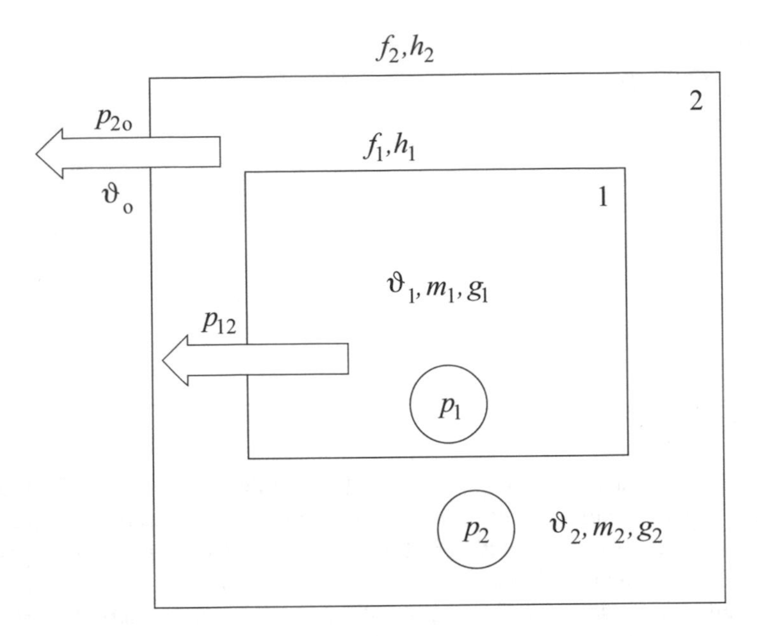

Fig. 16.1 A heat process

In the body of mass m_2 and specific heat g_2, power p_2 is converted to heat. This body encloses the body of mass m_1 and specific heat g_1, where the heat power is p_1. The surfaces where the two bodies are in contact with each other and with the external environment are f_1 and f_2, with heat transfer coefficients h_1 and h_2, respectively. Let us determine the change of temperatures ϑ_1 and ϑ_2 in the two bodies after switching on the heat generation, supposing that earlier the temperature of the system was equal to the environmental temperature ϑ_o. So the input signals of the system are the heating powers p_1 and p_2, the disturbance is the environmental temperature ϑ_o, and the output signals are ϑ_1 and ϑ_2, the temperatures of the two bodies.

The behaviour of the system can be described as follows. The temperature of the two bodies starts growing after switching on the heat. One part of the generated heat energy is stored in the heat capacity of the bodies, increasing their temperature, while the second part—as the effect of the temperature difference—leaves, entering the environment through the interfacial surface. It is supposed that the bodies are homogeneous, because of their good heat transfer properties a temperature difference does not take place inside the bodies.

In the inner body the heat generated over a time period of Δt partly increases the temperature of the body by $\Delta\vartheta_1$ degrees, and partly leaves for the outer body. The heat transfer depends on the difference of the temperatures in the two bodies, on the size of the interfacial surface, and on the heat transfer coefficient. In the outer body the heat transfer generated by the heat power p_2 is added to the amount of heat coming from the inner body. This resulting heat partly increases the temperature of the outer body by $\Delta\vartheta_2$ degrees, and partly goes into the environment.

The heating of the two bodies can be described by the continuity equations, $Q_{in} - Q_{out} = Q_{change}$, where Q denotes the quantity of heat. When heating a body this can be given as follows:

$$p\Delta t = mg\Delta\vartheta,$$

where p is the sum of the in- and out flow powers. $\Delta\vartheta$ is the temperature change in the body through Δt time, m is the mass of the body, and g is the specific heat. In the case of the two bodies

$$(p_1 - p_{12})\Delta t = m_1 g_1 \Delta\vartheta_1$$
$$(p_2 + p_{12} - p_{2o})\Delta t = m_2 g_2 \Delta\vartheta_2$$

where $p_{12} = h_1 f_1(\vartheta_1 - \vartheta_2)$ and $p_{2o} = h_1 f_1(\vartheta_2 - \vartheta_o)$. The heat flow between the two bodies depends on the temperature difference, the surface f between the two bodies, and the heat transfer coefficient h. Replacing the small changes Δ by differentials, the following equations are obtained:

$$\frac{d\vartheta_1}{dt} = -\frac{h_1 f_1}{m_1 g_1}\vartheta_1 + \frac{h_1 f_1}{m_1 g_1}\vartheta_2 + \frac{1}{m_1 g_1}p_1$$
$$\frac{d\vartheta_2}{dt} = \frac{h_1 f_1}{m_2 g_2}\vartheta_1 - \frac{h_1 f_1 + h_2 f_2}{m_2 g_2}\vartheta_2 + \frac{1}{m_2 g_2}p_2 + \frac{h_2 f_2}{m_2 g_2}\vartheta_o.$$

To simplify the equations, introduce the following notation (analogous to the notation in electrical systems):

$$R_1 = \frac{1}{h_1 f_1},\ R_2 = \frac{1}{h_2 f_2},\ G_1 = m_1 g_1,\ G_2 = m_2 g_2$$
$$\frac{d\vartheta_1}{dt} = -\frac{1}{G_1 R_1}\vartheta_1 + \frac{1}{G_1 R_1}\vartheta_2 + \frac{1}{G_1}p_1$$
$$\frac{d\vartheta_2}{dt} = \frac{1}{G_2 R_1}\vartheta_1 - \frac{1}{G_2}\frac{R_1 + R_2}{R_1 R_2}\vartheta_2 + \frac{1}{G_2}p_2 + \frac{1}{G_2 R_2}\vartheta_o$$

The input signals, output signals and state variables of the system are as follows.

$$\text{Inputs}: u_1 = p_1,\ u_2 = p_2,\ u_3 = \vartheta_0$$
$$\text{Outputs}: y_1 = \vartheta_1,\ y_2 = \vartheta_2$$
$$\text{State variables}: x_1 = \vartheta_1,\ x_2 = \vartheta_2$$

Choosing values for the parameters $R_1 = 1$, $R_2 = 1$, $G_1 = 1$ and $G_2 = 5$, the state equation of the system is

$$\dot{x}_1 = -x_1 + x_2 + u_1$$
$$\dot{x}_2 = 0.2x_1 - 0.4x_2 + 0.2u_2 + 0.3u_3$$
$$y_1 = x_1$$
$$y_2 = x_2$$

This is a system with three inputs and two outputs. A simplified case arises if only one input and one output is considered. Let $u = p_1$ be the input of the system (input heat power of the inner body) and let the output be $y = \vartheta_2$, the temperature of the outer body. Take the power p_2 as zero, so the outer body is not heated. The outer environmental temperature ϑ_o is considered as zero, which means the introduction of relative temperature values. With these assumptions the state equation of the system is

$$\begin{aligned} \dot{x}_1 &= -x_1 + x_2 + u \\ \dot{x}_2 &= 0.2x_1 - 0.4x_2 \\ y &= x_2 \end{aligned}$$

Analyse the behaviour of the system with MATLAB™.

$$\begin{bmatrix} \dot{x}_1 \\ \dot{x}_2 \end{bmatrix} = \begin{bmatrix} -1 & 1 \\ 0.2 & -0.4 \end{bmatrix} \begin{bmatrix} x_1 \\ x_2 \end{bmatrix} + \begin{bmatrix} 1 \\ 0 \end{bmatrix} u$$

$$y = \begin{bmatrix} 0 & 1 \end{bmatrix} \begin{bmatrix} x_1 \\ x_2 \end{bmatrix} + 0\ u$$

```
A=[-1, 1;0.2, -0.4]
B=[1; 0]
C=[0, 1]
D=0
```

The characteristic equation of the system is

$$\det(s\boldsymbol{I} - \boldsymbol{A}) = (s+1)(s+0.4) - 1 \cdot 0.2 = s^2 + 1.4s + 0.2.$$

The coefficients of this polynomial can be obtained also with the command poly.

```
karpol=poly(A)
    1.0000 1.4000 0.2000
```

The roots of this polynomial are the poles of the system, which are also the eigenvalues of $\boldsymbol{A}$.

```
p=roots(karpol)
        -1.2385
        -0.1615
p=eig(A)
```

The transfer function of the system is calculated by using $H(s) = \boldsymbol{C}(s\boldsymbol{I} - \boldsymbol{A})^{-1}\boldsymbol{B} + \boldsymbol{D}$. The polynomials that are its numerator and denominator can be obtained by the command ss2tf.

```
[num,den]=ss2tf(A,B,C,D)
      num = 0       -0.0000 0.2000
      den = 1.0000 1.4000 0.2000
```

or by using the *LTI sys* structure:

```
H=ss(A,B,C,D)
```

In transfer function form, this is

```
Htf=tf(H)
              0.2
      ----------------
       s^2 + 1.4 s + 0.2
```

In zero-pole form, this is

```
Hzpk=zpk(H)
              0.2
      -----------------
       (s+1.239)(s+0.1615)
```

Analyse the behaviour of the system in the time domain. The input signal is a unit step, $u(t) = 1(t)$ and calculate and plot the output signal (Fig. 16.2).

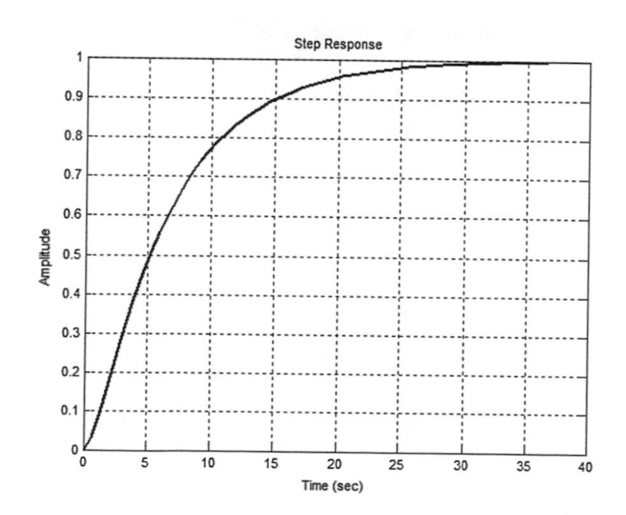

Fig. 16.2 Step response

```
step(H), grid
```

Give the output also in analytical form. The output can be calculated by using the LAPLACE transformation.

$$Y(s) = U(s)H(s)$$
$$y(t) = \mathcal{L}^{-1}\{U(s)H(s)\}$$

The LAPLACE transform of the unit step is $U(s) = 1/s$.

$$y(t) = \mathcal{L}^{-1}\left\{\frac{1}{s}\frac{0.2}{(s+1.239)\,(s+0.1615)}\right\}$$

It is known that

$$\frac{r}{s+p} \xrightarrow{\mathcal{L}^{-1}} re^{-pt}$$

The inverse LAPLACE transform can be found based on the partial fractional representation.

```
s=zpk('s')
Us=1/s
Ys=Us*Hzpk
```

Determine $Y(s)$ in polynomial form.

```
[num,den]=tfdata(Ys,'v')
```

Expand in partial fractions:

```
[r,p,k]=residue(num,den)
```

```
r =    0.1499
      -1.1499
       1.0000
p =   -1.2385
      -0.1615
       0
k =    []
```

In the case of single roots,

$$Y(s) = \frac{r(1)}{s-p(1)} + \frac{r(2)}{s-p(2)} + \frac{r(3)}{s-p(3)} + k = \frac{0.1499}{s+1.2385} - \frac{-1.1499}{s+0.1615} + \frac{1}{s}$$

The output signal in the time domain is

$$y(t) = 0.1499\mathrm{e}^{-1.2385t} - 1.1499t\mathrm{e}^{-0.1615t} + 1(t) \text{ for } t \geq 0.$$

It can be seen that the output signal resulting as a response to the input signal consists of two parts, a transient and a stationary part.

The first two components give the transient response which depends on the poles of the system. The third component gives the stationary response which depends on the poles of the input signal. Compare the outputs calculated in the two different ways.

```
t=0:0.05:40;
y1=step(Hzpk,t);
y2=0.1499*exp(-1.2385*t)-1.1499*exp(-0.1615*t)+1;
plot(t,y1,'b',t,y2,'r'),grid
```

In Fig. 16.3 it can be seen that the two curves coincide. Analyse the behaviour of the system for the input signal $u(t) = t^2/2,\ t \geq 0$. Its LAPLACE transform is $U(s) = 1/s^3$. Determine the stationary and transient components of the output signal. Give an analytical expression for the output signal.

```
Us=1/(s^3)
Ys=Us*Hzpk
[num,den]=tfdata(Ys,'v')
[r,p,k]=residue(num,den)
```

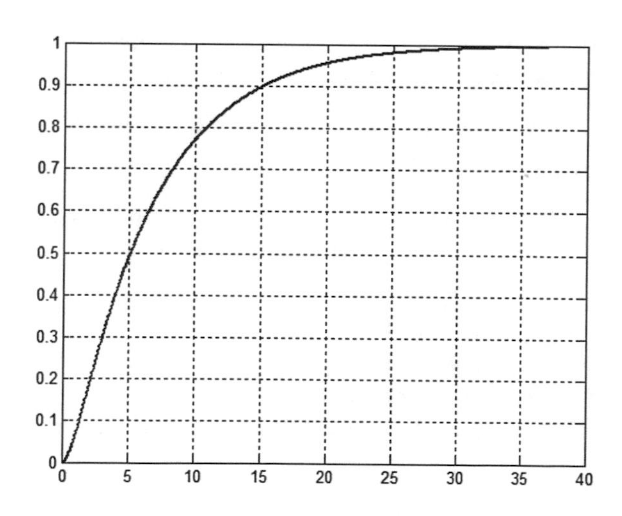

Fig. 16.3 Step response calculated in two ways

```
r =      0.0977
       -44.0977
        44.0000
        -7.0000
         1.0000
p =     -1.2385
         0.1615
         0
         0
         0
k =      []
```

The pole at $p = 0$ is a multiple pole of the system, therefore all of its powers have to be taken into consideration in the LAPLACE transform of the output signal.

$$Y(s) = X_2(s) = \frac{0.0977}{s+1.2385} - \frac{44.0977}{s+0.1615} + \frac{44}{s} - \frac{7}{s^2} + \frac{1}{s^3}$$
$$y(t) = x_2(t) = 0.0977e^{-1.2385t} - 44.0977e^{-0.1615t} + 44 - 7t + 0.5t^2$$

It can be seen that the first two components depend on the system dynamics (the poles of the system), and the last three components depend on the excitation. Separate these two components and plot them.

$$Y_{\text{tranz}}(s) = \frac{0.0977}{s+1.2385} - \frac{44.0977}{s+0.1615} \quad ; \quad Y_{\text{stac}}(s) = \frac{44}{s} - \frac{7}{s^2} + \frac{1}{s^3}$$

```
Ytranz=r(1)/(s-p(1))+r(2)/(s-p(2))
Ystac=r(3)/s+r(4)/(s*s)+r(5)/(s*s*s)
```

The entire signal is obtained as the sum of the two components

```
Ytranz+Ystac
```

The inverse LAPLACE transform can be found also with the impulse command, as the LAPLACE transform of the DIRAC delta is 1.

```
t=0:0.05:20;
y=impulse(Ys,t);
yt=impulse(Ytranz,t);
ys=impulse(Ystac,t);
plot(t,y,'r',t,yt,'b',t,ys,'g',t,yt+ys,'k')
```

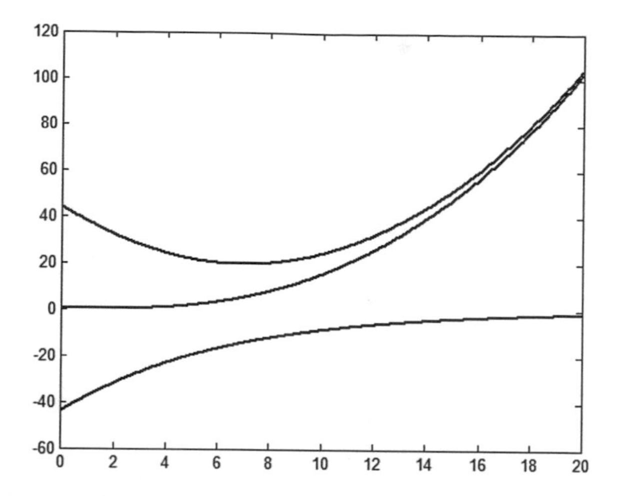

Fig. 16.4 The output signal is the sum of the stationary and transient responses

In Fig. 16.4 it can be seen that $y_{\mathrm{tranz}}(t)$ is decreasing and finally reaches zero (blue curve). The course of the stationary curve $y_{\mathrm{stac}}(t)$ depends on the input signal (green curve). The output signal $y(t)$ (red curve) is obtained as the sum of these two components.

Analyse the behaviour of the system for initial conditions. The system has two state variables. Set their initial values to be 10 and −10.

```
x0=[10, -10]
[y,t,x]=initial(H,x0);
```

The variable `y` contains the values of the output signal, `x` contains the state variables. Check the form of the state variables.

```
x
```

Separate the two columns. The colon: indicates all rows of the vector.

```
x1=x(:,1)
x2=x(:,2)
```

Plot the state trajectory.

```
plot(x1,x2),grid
```

Fig. 16.5 State trajectory

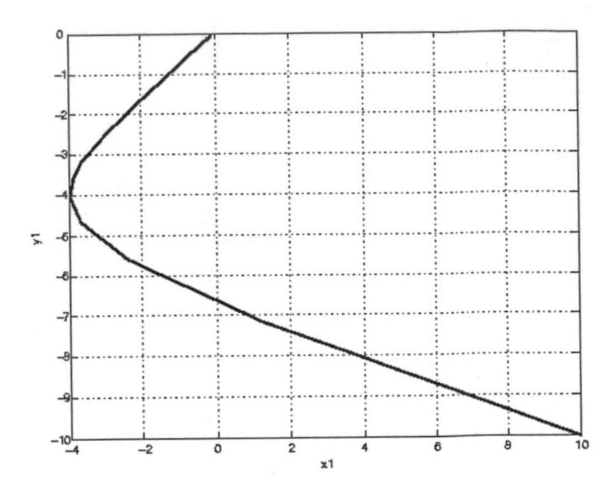

In Fig. 16.5 it is seen that the curve starts at (10, −10), the temperature increases in the colder body, while decreasing in the warmer one, then after the transients decay, both signals settle to the outer temperature, which is zero.

Problem Section 2.6 of the textbook [1] derives the models of a DC motor, of a tank and of the inverted pendulum. Simulate these models using MATLAB™.

References

1. Keviczky, L., R. Bars, J. Hetthéssy, and Cs. Bányász. 2018. *Control engineering*. Berlin: Springer.
2. Keviczky, L., R. Bars, J. Hetthéssy, A. Barta and Cs. Bányász. 2006. *Szabályozástechnika*. Budapest: Műegyetemi Kiadó, 2009.
3. Csáki, F. 1966. *Szabályozások dinamikája*. Budapest: Akadémiai Kiadó.
4. Csáki, F. 1973. *State-space methods for control systems*. Budapest: Akadémiai Kiadó.
5. Tuschák, R. 1994. *Szabályozástechnika*. Budapest: Műegyetemi Kiadó.
6. Åström, K.J., and B. Wittenmark. 1984. *Computer controlled systems. Theory and design*. Englewood Cliffs: Prentice Hall.
7. Åström, K.J. and T. Hägglund. 2006. *Advanced PID Control*. ISA—Instrumentation, Systems and Automation Society.
8. Åström, K.J., and R.M. Murray. 2008. *Feedback systems: An introduction for scientists and engineers*. Princeton: Princeton University Press.

L. Keviczky et al., *Control Engineering: MATLAB Exercises*,
Advanced Textbooks in Control and Signal Processing,
https://doi.org/10.1007/978-981-10-8321-1

Printed in the United States
By Bookmasters